December 1999

Margot

To a lady who has a passion
for fine vintage white linens

Enjoy

Martha L. Manchester

#216/10,000

Vintage White Linens

Marsha L. Manchester

Schiffer Publishing Ltd

4880 Lower Valley Road, Atglen, PA 19310

Dedication

To my dear grandmother, Lillian Hammond Burgess, and to my mother, Barbara Ellsworth Burgess Viola.

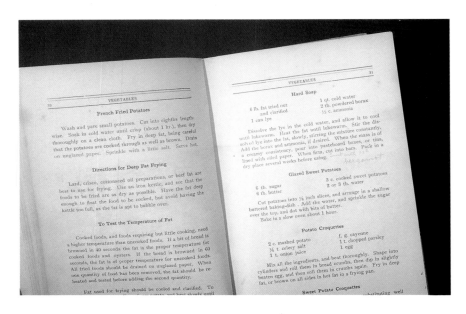

Manchester, Marsha L.
    Vintage white linen / Marsha L. Manchester.
        p. cm.
    Includes bibliographical references.
    ISBN 0-7643-0363-5
    1. Household linens -- Collectors and collecting
-- United States -- Catalogs. I. Title.
    NK8904.M36   1997
    746'.075--dc21   97-19734
                                    CIP

Published by Schiffer Publishing Ltd.
4880 Lower Valley Road
Atglen, PA 19310
Phone: (610) 593-1777; Fax: (610) 593-2002
E-mail: schifferbk@aol.com
Please write for a free catalog.
This book may be purchased from the publisher.
Please include $3.95 for shipping.
Try your bookstore first.

We are interested in hearing from authors with book ideas on related subjects.

Book Design by: Blair R.C. Loughrey

ISBN: 0-7643-0363-5
Printed in Hong Kong
1 2 3 4

Acknowledgments

To Christina Reneé Maury, for your friendship, support, and encouragement to make this book a reality. To Grace Dalton, for introducing me to your daughter, now my dear friend, Rose Ewas. If not for Grace and Rose's love of linens, the three of us would have never met. To Dee Rizzuto, for putting me up on the road to Atglen. To Susan M. Traversy, who has remained my dear friend through many years of business dealings. To Marjorie Levitt, a new found friend, for trusting me with her personal collection of wonderful textiles. To Susan Curran McCahon, whose spirit has inspired me through the years with her *joie de vivre*, and who shared with me her knowledge of Paris' *marché aux puces* shopping. To Jeanne B. Clarke, for her generous loan, for which I am most thankful. To Alan Miller and James, for sharing their enthusiasm of the linen business. To Whittier T. Brown and Barry L. Aldrich, for their love of fine linen. To Mimi Duphily, for sharing her knowledge and expertise, and for loaning part of her collection to be photographed. To Michael Mary Cowan, for her perennial friendship and for serving as my mentor. To Lori Vareika, for providing answers to all my questions. And finally, to Jessie Turbayne, for your relentless encouragement and your trust in my ability to accomplish this book. To all of you, I give my heartfelt thanks.

Upon stepping into the grange hall, it was love at first sight. The auctioneer was telling me to sit anywhere, but I was in such awe of all the wonderful "fresh from an estate" furnishings that I didn't hear a word he said. When I finally came down from cloud nine, embarrassed by having interrupted the proceedings, I selected seat number three, second row, left side - and kept it for the next five years.

At this first auction, I was transcended into a vast world previously unknown to me. The bidders numbered only twenty-five to thirty each week, and all were antiques dealers. Rarely did a retail customer like me show up.

With about ten items left to be sold (and only $25.00 in my pocket), I spotted a walnut, marble-top, half commode, a match to the one my grandmother had given me for a wedding present. Ever so shyly, I raised my hand to open the bid at $10.00. Eventually, I won the bid for $25.00, at $2.50 increments. Wow, I thought, did I get a bargain or did I pay too much, having just outbid everyone else in the room? My heart was pounding . . . surely everyone in the hall could hear it!

I took home this wonderful treasure to show my grandmother, and she agreed that it was a very close match to the other indeed. For the first time, I opened the commode drawer and door, only to find it full of the most exquisite European bed linens ever made, all tied together with faded pink silk ribbons. I removed the terribly soiled linens with an apprehensive forefinger and thumb, saying, "I'll get a rubbish bag to discard these filthy items." "No," replied my grandmother, "we'll make a batch of homemade soap and launder them."

Now, in 1970 I was the proud owner of a high speed, super deluxe, Maytag washing machine, and lived in a world of polyester wash and wear, so why - at twenty-two years of age - would I ever want to *make* soap when there was heavy duty Wisk and Clorox bleach to buy? Nonetheless, Grandmother took down her 1919 Brockton, Massachusetts, High School recipe book and looked up the recipe for hard soap and potato starch.

So...after three weeks of curing the pig fat and lye soap, came the use of an electric wringer washing machine to wash the linens, the peeling of 20 pounds of potatoes to starch the linens, and finally, the 5 pound dry iron used on a terry cloth padded, wooden ironing board to iron the linens. I thought I must have been really bad as a child and this penance was surely my pay back time! Yet as I lifted the exquisitely executed hand cutwork and lace embellished linen pillowcase off the ironing board, I knew then and there that white vintage linens were my passion.

That was how it all began. My mother and I opened an antiques store featuring refinished furniture, country collectibles, and linens—the first in the area. I left my forty hour a week job at the county courthouse and traveled to out-of-the-way country auctions in New England in search of forgotten treasures. I was actually making a living, and simultaneously loving the work. Life was sweet! At that time, there were no price guides available, and cartons full of yellowed vintage linens were truly a dime a dozen. Polyester was the rage, and women everywhere were disowning their irons.

I'd come home from auctions with blanket chests that were brimming with vintage heirloom linens. But how do you price these handmade treasures, not really knowing the difference between linen and cotton, between handmade and machine-made, between white and ivory? Who made the cutwork? Was it a nun in a convent, a young betrothed woman, a mother, or perhaps a grandmother preparing a trousseau for her granddaughter? Where do linen and cotton come from? There were no texts available to answer these questions, so I just kept stumbling along, learning more and more along the way.

Through this book, I want to share my passion for collecting, using, and enjoying vintage white linens. I want others to discover just how luxuriously these affordable treasures can change their everyday lives. If you've ever thought about collecting, now is the time to start, because there are still many homes with attics, hope chests, and steamer trunks filled to the top with linens. Every household has some form of linens, whether they are towels, sheets, pillowcases, or tablecloths. Many are beautiful items stored away for that special day that never came, and thus have never been used.

Just a few decades ago, in fact, wonderful lots of fancy linens were often relegated to cardboard boxes on the floor of an antiques dealers' shop or booth at most shows. Only in this decade have vintage linens become readily available, as the demand for them has increased. Even if a pair of pillowcases or some antique hand towels have not been passed down as family heirlooms, they still carry a time-worn softness of sentiment from cherished usage.

Today, buyers and collectors of vintage linens question the suitability of actually using their timeless treasures. Devotees combine vintage linens with other home decorations, using linens to underscore their good taste and style. Collectors who purchase freshly washed and ironed pieces feel an instant gratification, knowing they can take their purchase home and use it right away on either their table or their bed.

Although many books are now available on linens, few are devoted to the middle market everyday household linens still useful and affordable, and few contain a common sense price guide. The linens photographed for this book were all common in the past, but can still be found and are affordable today.

As a collector, you should become knowledgeable about the quality of the fabric, the location of the manufacturer, and the approximate age of the item. Once all these factors are taken into consideration, a price structure can be formed. Even though thousands of different linen styles exist, another example can always be found for comparison when pricing. The information in this book will primarily help the beginning collector to identify and price vintage linens from 1900-1950. Buy only what you like; if you pay too much for it today, by next year it will be worth the price you paid and you'll be ever so glad to have it!

Alençon - A needle lace in which fine designs of flowers and leaves are made on a delicate mesh ground, outlined with cordonnet, and linked with brides without picots. Most examples found today are machine-made lace.

Alençon Lace

A machine-made French Alençon white lace placemat from a set of eight matching placemats and eight matching napkins. It measures 16"L x 10"W. c. 1930s. $125-150.

This set of French ivory linen and Alençon machine-made lace consists of one dozen 24" square napkins. The matching tablecloth was beyond repair but the napkins were never used and retain their original label. c. 1920s. $75-100.

A French Alençon white lace runner in the Iris pattern. This scalloped edged runner measures 52"L x 20"W and is machine-made. c. 1930. $45-55.

Antimacassar - A three-piece set used on upholstered chairs to protect the two arms from hand soil and the back from the heavy oil men used on their hair. It was easier to wash these crocheted linen "tidies" than to wash the upholstery.

Antimacassar

A traditional fine English filet crochet lace sofa back in cream cotton thread with tassels, measuring 30"L x 12"W. This piece can also be used as a door or window treatment. c. 1941. $30-40.

Made of net filet lace, cutwork on linen, and bobbin lace, this 1920s French tidy is especially wonderful because of the tassels and the woman's profile in the center of the net filet lace wreath. $50-60.

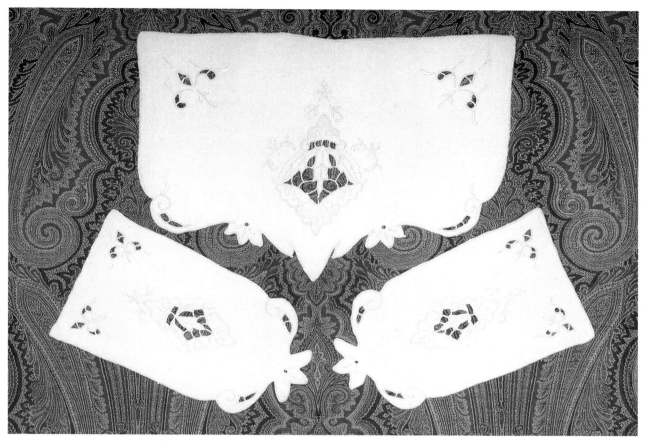

A typical Madeira three-piece linen chair set consisting of two arm covers and one chair back cover. This set has cutwork and light blue thread used for the embroidery. c. 1940s. $10-15.

Made in the 1940s of ivory French Normandy Lace, this chair back doesn't have any matching arm covers. Because this lace is rather fragile, the two arm covers probably wore out sooner than the head cover. $25-35.

A 1920s three-piece Italian white linen chair set with cutwork, rose filet
lace edging, embroidery, and reticella needle lace insertions. The head
cover measures 14"L x 9"W and the two arm covers are 7"L x 9"W. $45-55.

Appenzell - A Swiss made, cottage industry type of padded satin stitch white-on-white embroidery mostly found on linen. This type of embroidery is probably the finest white-on-white work made.

Appenzell

An Appenzell three-piece linen towel set with two hand towels and one bath towel. Appenzell work was copied by many people other than the Swiss; as a result some good copies can be confused with the real work. The woman from whom I purchased this set received it as a trousseau gift from her godmother in 1922. $100-125.

Like the Appenzell towel set, these fine linen pillowcases and queen size top sheet were part of a woman's trousseau in 1922. They were used for one week only during her honeymoon and stored away ever since. $225-250.

A closer look at the linen Appenzell pillowcases shows wonderful detailing in the tiny hand embroidered stitches. Because this sheet set was washed only once in its lifetime, the grey penciled stencil marks have not washed off. In due time, the pencil marks will disappear after repeated washings.

Appliqué - A method of applying lace, fabric, a pattern, or an image onto an existing woven fabric.

Appliqué

This pair of matching bureau scarfs was made in Madeira. The linen pastel appliqués are hand stitched onto the white linen fabric. The long scarf measures 41"L x 17"W and the short scarf is 38"L x 17"W. c. 1940s. The set is valued at $45-55.

Purchased in Paris, this silk scarf with cotton appliqués sewn on by hand is from Belgium. The ivory silk is in perfect condition, considering its age of 95-100 years. 80"L x 22"W. $75-100.

Another Paris flea market purchase, this white, hand appliquéd
on net scarf measures 78"L x 22"W. c. 1910. $75-100.

Opposite page bottom:
A fine example of machine-made net filet lace with
alternating squares of hand-sewn mosaic linen squares.
This tablecloth looks outstanding when covering a dark
wood table. 100"L x 68"W. c. 1940s. $100-125.

Army-Navy - A name used primarily for tablecloths and runners which consist of alternating fabric and lace squares. The fabric is either linen or cotton with machine-made or hand embroidered padded stitches. These items were made in China mainly for American servicemen to bring home as gifts after World War II.

This Army-Navy linen and lace piece measures 48"L x 18"W and can be used either as a bureau scarf or table runner. c. 1940s. $25-35.

Art Deco - With the advent of the flapper era, people tired of stark whiteness and desired a style revolution featuring straight, clean lines, and soft, silky, lightweight fabrics in pastel and brighter colors. Following the discovery of King Tutankhamen's tomb, Egyptian motifs came into vogue. New trade agreements between the United States and Japan brought a plethora of Oriental influenced linens from 1920-1940.

Art Deco

A wonderful example of silk embroidery on flax colored linen. The ribbon effect is carried all around the outside edge of the doily. The vibrant yet natural colors of the violets are so lifelike. 24" round. $25-35. *Collection of the author.*

Close-up of the fine embroidered flowers. Note the light hand painted green edge on the ribbon.

A most unusual find. The vivid silk hand embroidered designs on this rare
collar are truly remarkable for the collar's age and usage. Apparel of this sort
is not usually intact, and, if found, has been washed so often that the original
colors have been lost. c. 1930s. $75-100. *Courtesy of Marjorie Levitt.*

Art Nouveau - An innovation of design in fine arts and architecture which began in the late nineteenth century and lasted until World War I. As the Industrial Revolution created machine-made products it became both fashionable and affordable to lavishly decorate one's home with linens galore. The most characteristic Art Nouveau motifs are sinuous and curvy lines, nudes, semi nudes in human form, and women with flowing hair, all evoking exotic designs.

Art Nouveau

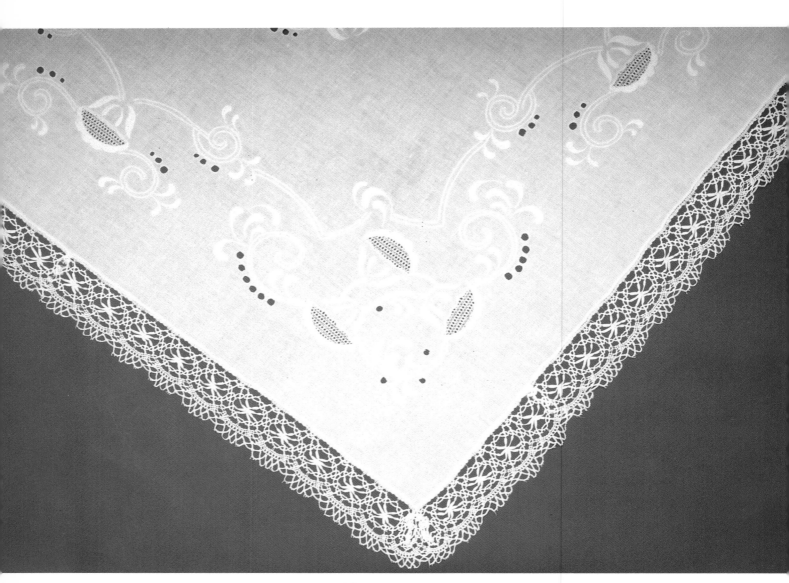

An Art Nouveau 68" square linen tablecloth with stylized flowers in padded satin stitch, handmade embroidery, needle lace inserts, and a needle lace border. $65-85.

Arts and Crafts - The Arts and Crafts Society began its movement in England in 1888. The Society's objective was to create a new range of textiles based on the techniques and traditions of the past. Vegetable dyeing and the art of hand block printing became highly successful. The American Arts and Crafts movement, founded nine years earlier by Associated Artists, used the same basic principles as the English Arts and Crafts movement. Arts and Crafts designs are generally abstract or feature naturalistic, stylized themes, such as flowers. The linen fabric used was a coarser, heavier weave and darker in color. Colors of the embroidery thread were much more vivid and bold over muted tones of fabric.

Arts and Crafts

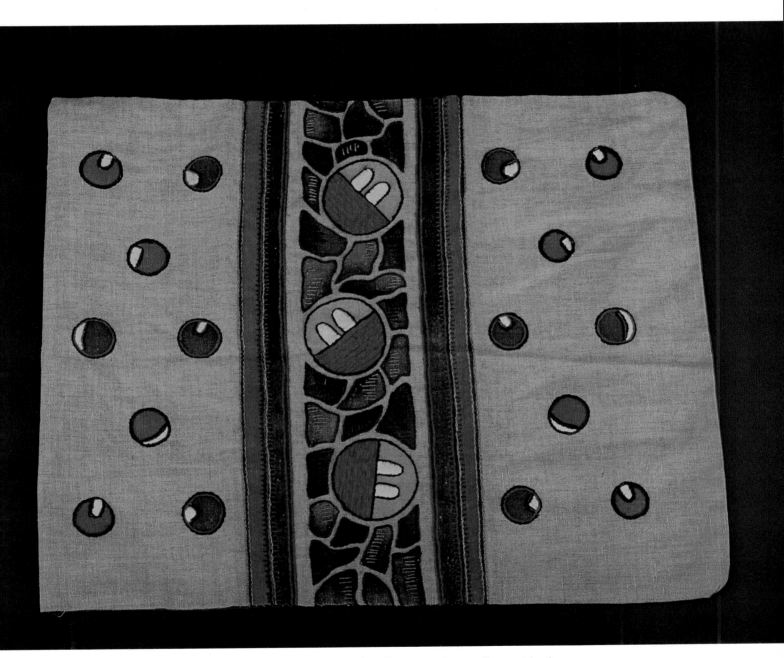

A truly wonderful Arts and Crafts pillow cover in ecru linen with vibrant colors and stylized designs. 20"L x 16"W. $65-85. *Courtesy of Marjorie Levitt.*

An Arts and Crafts pillow cover in ecru linen with handpainted flowers, leaves, and butterflies all outlined in embroidery floss. 20"L x 14"W. $25-35.

A closeup of an Arts and Crafts large round doily with beautifully executed French knots. *Courtesy of Marjorie Levitt.*

Auctions - At one auction I attended, I overheard some ladies whisper that whenever I bid, they would bid one bid higher because they figured I got linens for practically nothing and they didn't have overhead expenses like I did. Then, when I was successful that day in bidding for a box lot, I wondered whether I overpaid for the lot since I was willing to pay more than the other 200 people at that auction!

The key with auctions is to know your auction house, question whether or not that box of fancy linens truly did come from an estate (or whether a dealer just consigned his or her shopworn linens to the auction), and to remember that auction houses are in business to make money, not to give anything away or sell it too cheaply. Before the auction, take the time to preview every item in which you are interested. Carefully inspect the item's condition, measure its size, check the quality, and set a firm price in your mind. Everyone gets caught up in the pace of the auction, and it's difficult to make quick decisions during the actual bidding. So stand firm in your set price and you won't have any regrets later about overbidding.

Baby Pillowcases - Small rectangular feather pillows about 16" long x 12" wide were used mainly for sleeping. The pillowcases or pillowslips for these pillows have an opening on one end only.

Baby Pillowcases

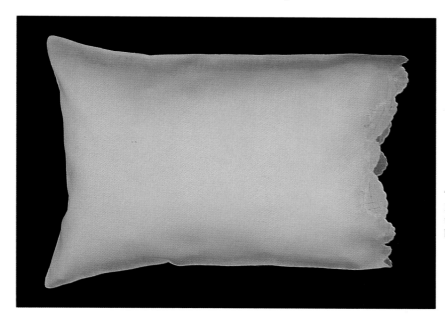

This baby pillowcase, made in the Philippines, is of fine cotton with the tiniest hand stitched embroidery so befitting a baby's pillowcase. c. 1940s. 16"L x 12"W. $20-30.

A baby pillowcase decorated with a bunny, chicks, and flowers. This pillow cover has a button back closure and is outlined in hand stitches. It was made exclusively for B. Altman Company, New York and the label attached is from Switzerland. This organdy cover measures 16"L x 11"W. c. 1950s. $20-30.

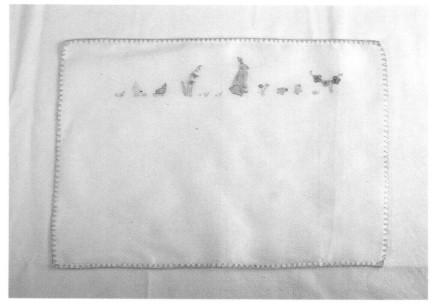

Baby Sheets - The majority of baby sheets made were of high quality cotton percale or linen, and usually had a decoration of cutwork, embroidery, lace and/or appliqués.

Baby Sheets

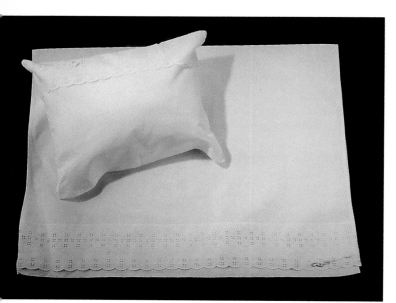

A Paris yard sale find, this cotton baby top sheet and pillowcase is embellished with machine-made eyelet embroidery (Broderie Anglais). c. 1950s. $35-45.

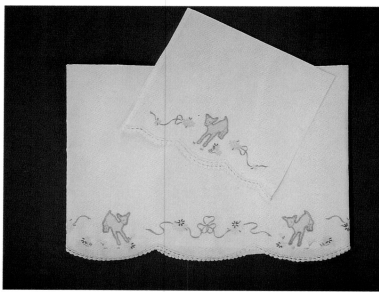

A baby top sheet and matching pillowcase with appliquéd lambs, hand embroidery, and hand crocheted lace on a scalloped edge. c. 1920s. $35-45.

An Italian cotton baby top sheet with hand cutwork, net filet insert, and embroidery. From a Boston estate, this sheet measures 60"L x 40"W. c. 1920s. $25-35.

Baby Shams - To show off a new baby in its baby carriage, small rectangular feather pillows were used to prop up the baby's head. The pillowcovers used for these pillows were usually fancier than regular bed pillowcases, and were embellished with lace, ruffles, cutwork, appliqués, and embroidery. These shams came in three styles: buttonhole closure on the back, an envelope style flap on the back, or buttons on both ends. Since carriages are rarely used today, baby pillow shams are now used in bedrooms as boudoir pillows.

Baby Shams

A very fine example of a French linen baby sham with hand appliqués, drawnwork, minute embroidery, and a scalloped edge. c. 1920s. $30-40.

A closer look at the back of the French linen baby sham shows a very rare find: the strip of fabric with the attached buttons closes the back of the sham.

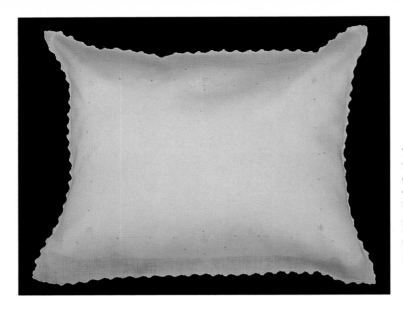

This fine cotton baby sham with soft pink and blue hand embroidery has a button closure on the back and retains its original feather pillow inside. The pillow has had some new feathers added and the old ones were plumped up. c. 1940s. $25-35.

Battenberg Lace - The name of this lacework is often mistakenly applied to many other forms of lacework. The general form of true Battenberg lace is machine-made tapes basted down on a pre-stamped muslin pattern. The laceworker then fills in the empty spaces with fancy stitches. Early examples were always made with fine linen centers and the best spider work. When printed patterns from stores and kits ordered from catalogues became available, the fabric centers were made of muslin and the lace fillings were made of coarse threads.

Battenberg Lace

This 48" round German tape lace doily is a fine example
of better quality craftsmanship. c. 1920. $125-145.

Opposite page:
This grape and leaf design was one of the most used patterns
in the commercial production of do-it-yourself home kits of
Battenberg Lace around 1900. Round cloths such as this came
in sizes as small as 6" and as large as 90". The larger cloths are
harder to find and almost always need repair. This 72" round
tape lace cloth has a cotton center. c. 1900. $75-100.

Battenberg Lace on net curtains in pairs of two or more is very difficult to find. If the sun and wind didn't rot the fibers during their lifetime, modern chemical detergents and high speed washing machines would do the deadly deed. These date from 1920-1930 and are 90"L x 32"W. $100-125 for the pair.

I believe if you placed every doily ever made end to end, they would circle the world ten times! A most useful and practical item, the doily should always be pretty. Small doilies like this often came in sets. This Battenberg tape lace doily is 10" square with a cotton fabric center. c. 1920s. $10-15.

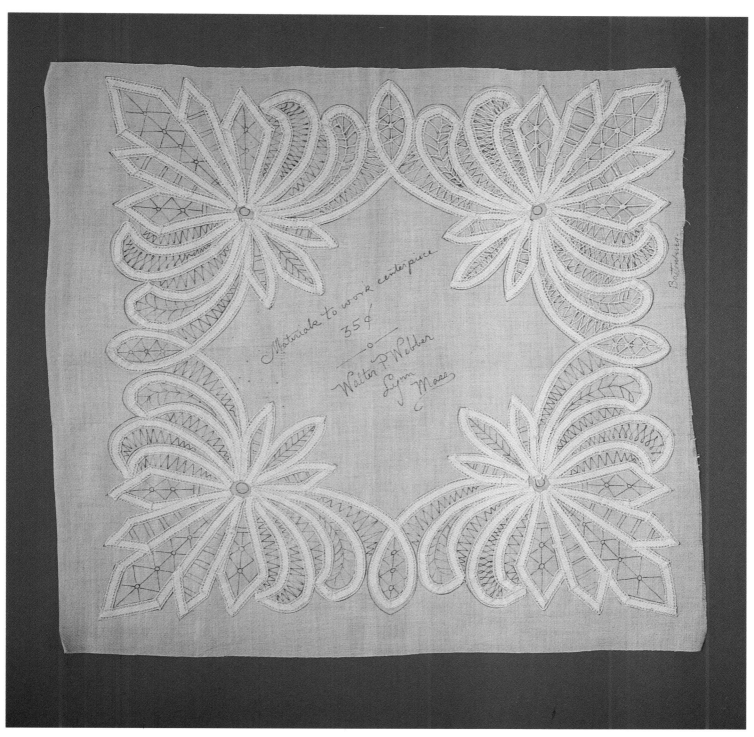

From the Walter P. Webber dry goods store in Lynn, Massachusetts, this Battenberg tape lace doily pattern has a beginning but no end. The tapes are basted on the pink muslin and some of the handwork was started but never finished. It measures 24" square and dates from 1918. $20-25.

A white tape lace runner found in Europe that has some very
fine executed stitches. It measures 54"L x 19"W. c. 1920. $65-75.

Beaded Embroidery - Not a true form of embroidery, this not so popular handwork is found only occasionally. At the height of beaded handwork popularity, Native Americans were making beautiful examples of other beaded handwork items, but not necessarily beaded embroidery. Most Native American handwork was available for sale along the roadside, mainly for the tourist trade.

Beaded Embroidery

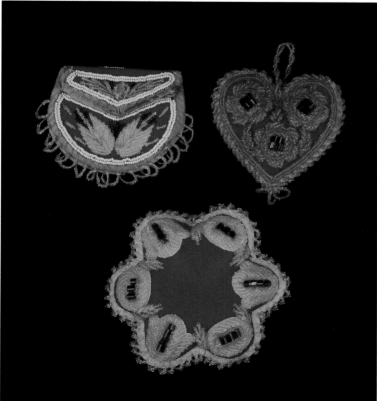

Above:
Not a very easy piece to iron, this glass beaded runner is made on a poor quality muslin fabric with a machine-made lace edge. It measures 46"L x 18"W. c. 1920s. Because very few pieces of glass-beaded embroidery exist, the value is $35-45.

Left:
A few examples of Native American beaded work that were sold on the roadside in New England. c. 1920-1940. $45-75 each.

Bed Covers - Bed covers were used not only for decoration, but also for protecting the wool blanket. Bed covers, bedspreads, and bed coverlets range in fabric from heavy, coarse muslin to fine, sheer linen and lace. Like sheets from 1900, bed covers were very oversized. Those large covers now fit our modern king and queen size beds, and some open work bed covers make fine canopy tops. Bed covers come in many different varieties:

Chenille spreads (the name is derived from the French word for caterpillar), were cotton tufted spreads popular in the 1930s through the 1950s.

Cotton coverlets, as the word implies, were coverlets that were meant to come just below the mattress; a dust ruffle was needed from the innerspring to the floor to keep drafts and dust from getting under the bed.

The ever popular *cotton crocheted bedspread* was made during the Great Depression; women were home more often since any available jobs were given to men. As bedspreads, these bed covers went all the way to the floor.

Duvets, of European decent, are covers that have a separate top and bottom. Three sides are stitched together while the fourth has a button closure. A feather filled comforter, called an eiderdown, is inserted inside the duvet before it is buttoned shut (it is easier to wash a linen or cotton duvet cover than to constantly dry clean the feather filled comforter). Most vintage European duvets are available, but rarely do our modern American comforters fit inside. The matching pillowcases are European square cut, not rectangular like modern American pillows.

Most *linen bed covers* were made of linen from Ireland, the worlds' second largest grower of linen, and the worlds' first and foremost exporter of linen. The quality of Irish linen was unsurpassed, and consequently the most exquisite handwork was executed by highly skilled craftsmen.

Marseilles are cotton, machine-made, white-on-white coverlets that derive their name from the French Riviera town of their origin. The originals were stuffed with cotton to form raised patterns. Most trapunto Marseille coverlets found in New England that are "new, old stock" (i.e., ones never used and retaining their original paper labels) indicate they were made in Lewiston, Maine.

Normandy Lace bed covers are called "Patchwork" lace because of the various machine-made Paris Lace, handmade lace, and embroidered linen inserts. This lace has a fragile net background, and most used pieces found usually have some damage as a result.

White-on-white quilts are called brides' quilts. The white embroidery on white fabric represents purity.

Bed Covers

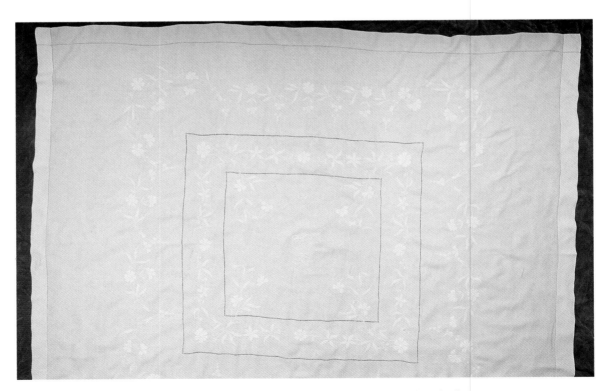

A linen bed cover from the United Kingdom. The tailored edge
has a row of hand drawnwork, machine embroidered flowers,
and a monogram in the center. 98"L x 94"W. c. 1920s. $85-100.

This Italian trousseau bed cover is a very fine example of hand-painted linens. The reason the colors are so bright is that it was never used and has been folded away since the 1920s. It measures 120"L x 110"W. $750-850.

A close up of the fine hand painted detail on the Italian trousseau bed cover.

The net mesh background of this bed cover is machine made while the two-tone needle work is handmade. The figural *puttis* (Italian angel-like figures) and the rose subject matter are finely made. It measures 100" square and fits a queen size bed. c. 1915. $200-250.

Often called the "All American Coverlet," this cotton, machine-made, scalloped edged, cut cornered, double bed sized coverlet always had a matching pillow sham that only lay over the top of the pillows. These easy care bed covers bring back memories of grandmother's upstairs cool bedrooms on hot summer nights. c. 1880-1940. $75-100.

Right:
A close-up of the detail on the Marseille type coverlet shown below.

Below:
A cotton "trapunto" style machine-made Marseille type coverlet. This was made in Lewiston, Maine, c. 1890, as indicated by the paper label it had before it was washed. It measures 84" square and fits a double bed. $150-200.

Bed Sheets - Bed sheets were a household necessity, and a young bride-to-be would usually acquire a minimum of one dozen for her trousseau. European royalty and the affluent lived in large stone castles, so their beds were also of a large size. Only the finest bed linens are large enough to fit our modern king and queen size beds, those sheets being 90" to 100" wide. Many times people actually mistake fancy bed sheets for tablecloths. The bed sheet has lace, embroidery, or cutwork across the top, and often the same decoration down both sides for about 10" to 20". This is where you fold over your top sheet to the blanket, thus protecting the precious blanket from wear.

Bed Sheets

This lovely French linen double bed top sheet has nasturtiums hand embroidered among hand netted lace. The matching one piece pillow bolster was never cut to make two pillowcases; it also has the monogram of the original owner, Marie Clariot, from Lyon, France. c. 1930. $200-250.

Opposite page:
This French ivory Normandy Lace bedspread has a fine line to distinguish the pillow tuck-under from the top of the spread, as well as a net lace embroidered ruffle. 104"L x 84"W, it fits a double bed. c. 1940s. $600-700.

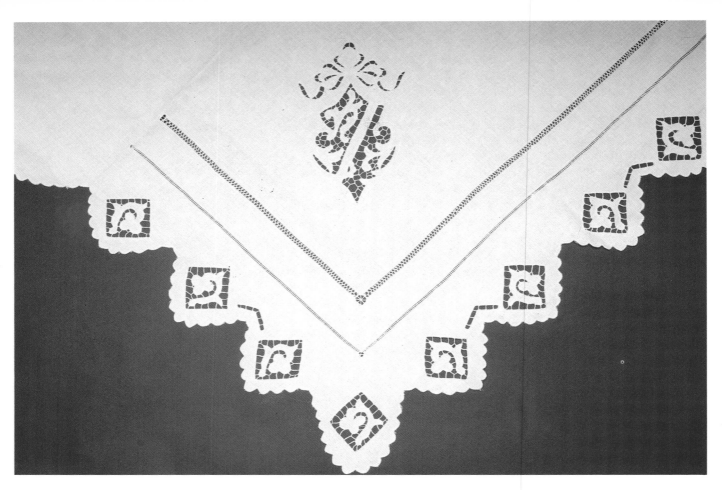

Above:
A queen sized Italian cutwork and drawnwork cotton top sheet with machine-stitched embroidery and a fancy scalloped edge. These sheets were sold in sets consisting of either a pair of pillowcases, a pair of pillow shams, or a lay-over-the-top sham. This complete set includes bureau scarfs and end table doilies. They were made for export to the United States and sold at moderate priced stores. The blanket fold over begins where the scalloped edge stops. The complete set is valued at $150-175. c. 1940.

Left:
A close up of the center detail on the Italian top sheet.

Bed Sizes - Standard bed sizes are: twin 39" x 75", full 54" x 75", queen 60" x 80", and king 78" x 80". This is the actual size of the top of the mattress.

Belgium Lace - First time lace collectors are sometimes under the impression they can find fantastic examples of fine, handmade Belgium Lace at flea markets and general stores in Belgium—not so! This exquisite lace is still being made, but lacemaking is very costly and time consuming, and very few apprentices are willing to spend their whole life's work just making lace. Due to the prohibitive cost, the Belgium Laces still being made are generally commissioned only to royalty. Fine examples of Belgium Lace are Brussels, Bruges, Mechlin, and Flanders; examples of these laces can usually be found in museums.

Blanket Covers - The European style is a flat top protector with buttonholes on three sides to fasten the cover to the blanket (which has the corresponding buttons on three sides). To protect the top of the blanket, the fourth side of the cover has a turn over with some sort of handwork. The American style of a blanket cover is two sheets sewn together on four sides, but with an opening in the very center about 24" to 30" round, just wide enough to slip the blanket into. The advent of satin binding on blankets sealed the fate of the blanket cover.

Belgium Lace

This set of cocktail napkins has its original paper label stating "Made in Belgium." The peach linen center has the Belgium lace edge made of ivory linen thread. The fitted box has silk ribbon to tie the napkins together. c. 1930s. The set of twelve cocktail napkins is valued at $60-75.

Blanket Covers

A cotton blanket cover that is 80"L x 70"W with an opening in the very center. The 24"W opening is used to show off what there is to see of the blanket. These were not used too often, thus very few exist. c. 1940. $45-55.

Bobbin Lace - This is lace made on a pillow by twisting and plaiting threads. The linen threads are first wound on bobbins. France, Spain, and Flanders (an area between Mechlin and Antwerp) all claim credit for creating bobbin lace, but most historians agree that the creators were Italian nuns. Forms of straight bobbin lace are Torchon, Cluny, Maltese, tape lace, and Chantilly. Forms of free bobbin lace are Honiton and Belgium Duchesse. Queen Victoria's wedding veil of 1840 was made of Honiton Lace, from Honiton, England.

Bobbin Lace

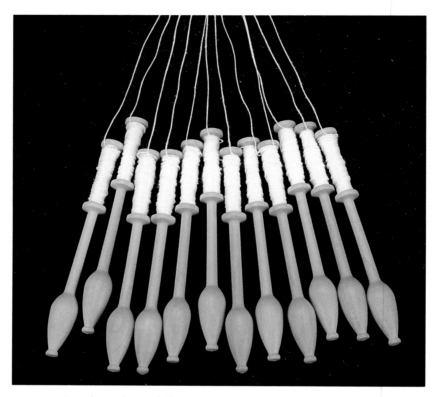

One dozen beautifully made wooden bobbins from England.
Linen thread is wound on them. *Courtesy of Mimi Duphily.*

A strip of handmade linen bobbin lace with the pattern for making
this type of lace underneath. Handmade by Mimi Duphily. c. 1996.

Bolster Cases - These are one piece pillowcases made for either one long round pillow or two rectangular pillows. Some textile mills gave the housewife a choice when purchasing the one piece bolster case: she could either keep it as one long case, or cut it in half to make two pillowcases. New bolster pillows are still available in Europe at department stores.

Bolster Cases

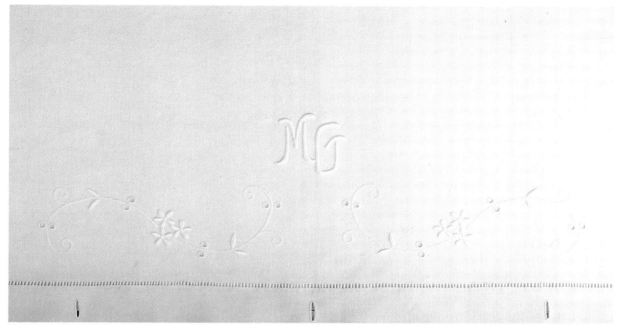

A German bolster cover with a monogram of "MG", padded satin stitched flowers, a little cutwork, a row of hemstitching, and a row of buttonholes with buttons (not shown) on the back side. This bolster case is for a double sized round pillow meant for two people to sleep on. c. 1910. 70"L x 40"W. $45-55.

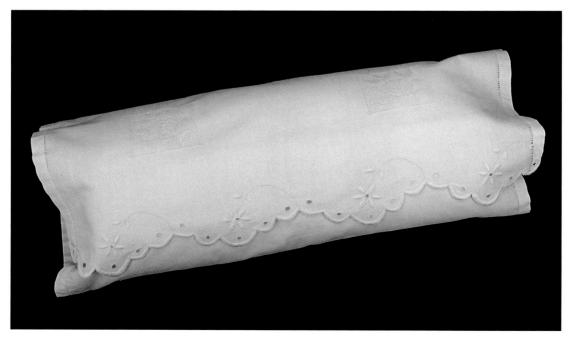

A single pillow bolster with embroidery, lace inserts, a little cutwork, and a scalloped edge. 18"L. c. 1940s. $25-35.

Boudoir Pillows - These fancy bedroom pillows were lavishly decorated with the finest laces, embroidery, and fabrics. They were used on the fainting couch for milady's afternoon nap. Some boudoir pillows started out as baby pillow shams.

Boudoir Pillows

Originally used as a baby carriage pillow, this Italian linen boudoir pillow cover has a *putti* of reticella lace, hand cutwork, hand embroidery, and a machine-made Rose Filet Lace edge with a button back to keep the pillow inside. It measures 17"L x 12"W. c. 1920s. $50-75.

Above:
A French heart-shaped boudoir pillow in linen with a monogram and a Rose Filet Lace edge. c. 1920s. $55-65.

Left:
A round, machine-made lace boudoir pillow with angels pulling chariots of bare-breasted women; a boudoir piece for sure. 12" round. c. 1940s. $55-65.

Bread Tray Mats - These were crocheted oval mats which were placed on top of the bread. Flour or corn meal was then sprinkled on top of the mat, leaving the impression of the crocheted word on the loaf. Linen bread tray mats were placed on the bottom of the tray, with the bread on top of the mat.

Bread Tray Mats

Right:
Top: A Madeira linen bread tray mat with light blue embroidery. c. 1920s. $12-15. Middle: A hand embroidered round linen bread tray mat, made from a kit. c. 1940. $10-12. Bottom: An oval hand embroidered linen bread tray mat, also made from a kit, c. 1940. $12-15.

Below:
An ecru crochet lace "Staff of Life" bread tray mat and a white crochet "Bread" mat perfect for framing if not used on the bottom of an oval bread tray. c. 1940s. $10-15 each.

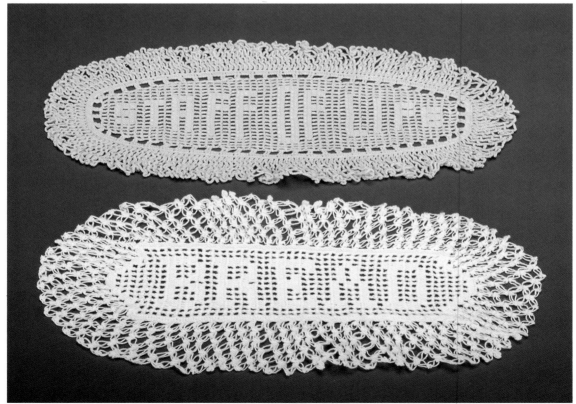

Bridge Set - The game of bridge became popular during the roaring twenties, thus necessitating the need for pretty table covers to place on the folding card tables. These 34" to 42" square cloths had matching napkins. Today we use them as table toppers to fit round or square side tables.

Buratto - A pattern that is tightly woven on a knotted net lace background, Buratto is from Italy and very similar to net darning or Rose Filet Lace. Very few examples are found today, probably because it was not a favorable lace to make.

Buratto

Two Buratto doilies on white linen. Although the patterns are both very appealing and very easily made, not many examples are found on the linen market. c. 1920s. Each $12-15.

Buttonhole Stitch - A simple, but very sturdy, slightly padded satin stitch made expressly for buttonholes. On most embroidered pieces, the buttonhole stitch was used for scalloped edges and for cutwork.

Buttonhole Stitch

This is a very fine example of a buttonhole stitched doily on white linen. Possibly made from a kit. c. 1915. 24" round. $25-35.

A close up of the fine detail of the buttonhole stitch.

Carrickmacross - From Carrickmacross, Ireland, this specialty lace has a muslin or linen fabric appliquéd to a fine net background with loop edges. The appliqué is held in place by cordonnets, a heavy outlining of threads. The Carrickmacross process is hardly a short one: a short bridal veil, custom ordered and still being today made by nuns, will take four to five *years* to be completed!

Chantilly Lace - First made in the second quarter of the eighteenth century in Chantilly, France, this was black lace made of flax on a mesh net background. This beautiful bobbin lace has a delicate floral pattern with ribbons and swags. During the third and fourth quarter of the nineteenth century, machines began mass producing large shawls and mantillas of Chantilly Lace. So honored was this symbol of a woman's dignity that her mantilla could not be confiscated for debts.

Opposite page top:
The Irish Carrickmacross white linen centered bridal handkerchief was made in 1922. 12" square. $75-95. The ivory collar is approximately the same vintage as the handkerchief. $45-55.

Opposite page bottom:
The black Chantilly Lace and the ivory lace piece are both women's scarfs. Each measures 80"L. These machine-made laces were manufactured around 1900. $45-55.

Chantilly Lace

Cleaning - There is almost nothing more sensual than the touch, sight, and smell of freshly washed, dried, and ironed linen, especially linen towels and bed sheets. To attain this experience, one must gently remove the soil of all those many years. The longer the soak, the cleaner the fabric. Most vintage linens were always washed in gently boiling water with old-fashioned soap (not detergents), and never agitated. For the past twenty-seven years, I have followed my grandmother's method for washing linens and have never lost as much as one piece to this day:

Fill your washing machine with tap water if you are washing whites. Add two ounces of Linen Wash by LeBlanc or a non-detergent soap, such as Ivory Snow. If any suds appear, scoop them out and discard them. Inspect the linens for any spots that need extra removal. Wet the spot with water, add a few drops of Linen Wash, and let soak for 15 minutes. Place all small and delicate pieces in a mesh bag, then drop the linens in the machine and by hand gently squeeze the soapy water through the fabric, making sure all pieces are completely wet. Let soak for at least six hours. This initial soaking of very soiled linens will loosen the dirt and grime very gently.

After the six hour soaking is finished, turn your washer dial to a gentle agitation of only ten seconds and empty the tub. Refill the tub with cold water and swish the linens by hand for 30 seconds, then empty the tub. Refill the tub again, and by hand swish the linens once more to be sure all soap is removed. Empty the tub and gently spin the linens for only ten seconds to remove the excess water.

Remember, these are sturdy everyday household linens that have survived for a lifetime or two because of proper washing methods. The results of your cleaning will be outstanding and well worth the trouble. Afterwards, your sense of sight, touch, and smell will transcend you into a world of wanderlust and you will wonder why you don't do this more often!

Cluny Lace - Originally from an antique house in Paris (now the Musée de Cluny), this simple bobbin-made, geometric design has slender rosettes, diamond blocks, and an edge of braided loops. Machine-made examples of Cluny Lace presently being produced worldwide are hard to distinguish from the original handmade laces.

Cluny Lace

A fine example of French Cluny Lace with a white linen fabric center. This doily is 30" round. c. 1920s. $45-55.

Above:
These two beautiful Cluny Lace and linen runners can be used as dining table runners or as bureau scarfs. They each measure 54"L x 19"W. c. 1920s. Each $75-95.

Right:
The ivory French Cluny Lace doily carries its original tag of origin so there is no question of its authenticity. The workbook shows two patterns of crocheted Cluny Lace to be made, whereas the French Lace is bobbin made by hand. c. 1920s. $25-30.

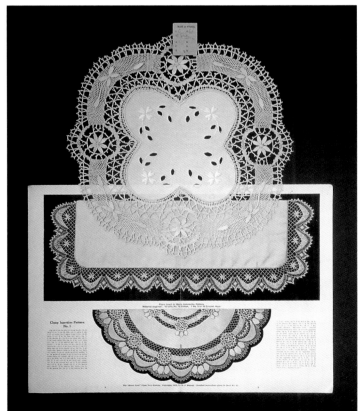

Coasters - Used as a "catch all" for spills or as an underplate doily, coasters came in sets of six and twelve. Cocktail glass coasters were to prevent any liquid spills from ruining lacquered furniture due to the white rings left behind from moist glasses. Fancy underplate coasters were to absorb any soup spillovers on the dinner plate. Today those paper lace doilies are available, but why use paper when you can have the real thing?

Coasters

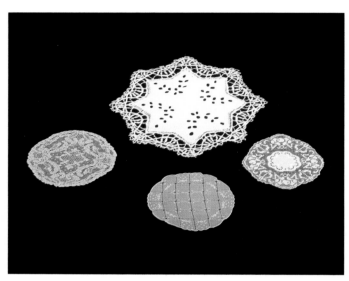

These four French Lace coasters were originally used as doilies placed between the cup and saucer to absorb any liquid in the saucer. Today, they are used as glass, undersaucer, and underplate coasters. c. 1920s. Set of twelve, $36-48.

Four Belgium Lace glass coasters. They were put on the bottom of stemware glasses to protect varnished furniture from scratches or water marks. c. 1950. $20-24.

A stemware glass with a glass coaster on the bottom, showing the characteristic rooster (or cock, for "cocktail") motif. c. 1940s. Set of six, $20-24.

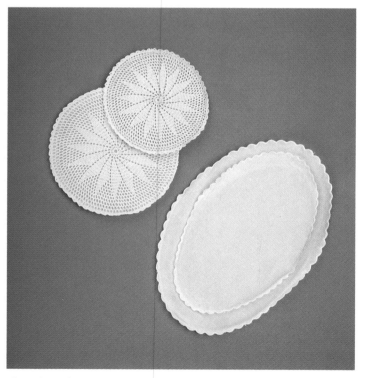

These hot plates, often called tureen coasters, have an asbestos liner underneath the two crocheted covers and inside the two oval linen covers. c. 1940s. Each set $15-20.

Cocktail Napkins - Like glass bottom coasters used for drinks, cocktail napkins were used for finger food such as appetizers. The most often used motif on these square or rectangular napkins was the rooster, or cock, thus the name cocktail.

Cocktail Napkins

A Madeira cocktail napkin and glass coaster set. These linen pieces are all hand embroidered. c. 1940s. $24-36.

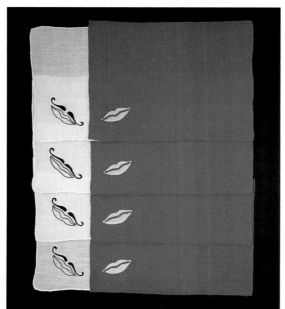

Eight Victor/Victoria linen cocktail napkins from Madeira. c. 1920s. $24-32. *Courtesy of Alan Miller and James.*

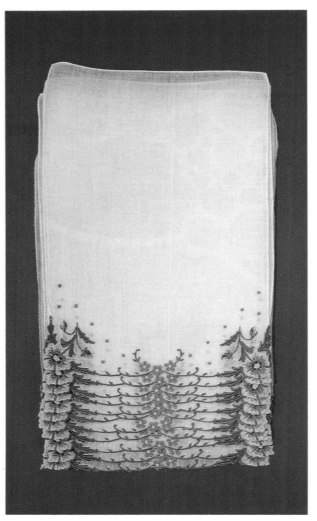

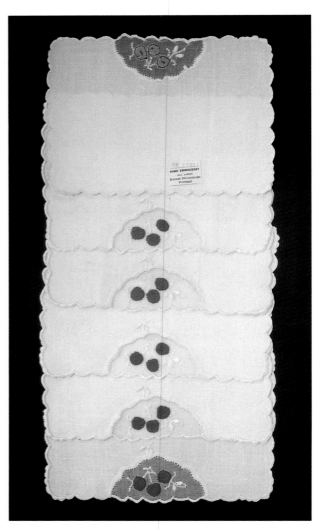

These are lovely Swiss embroidered cocktail napkins. c. 1930s. $36-48.

Linen cocktail napkins from Portugal with hand appliquéd cherries. Never used. c. 1950s. $36-42.

This set consists of eight cocktail napkins and eight glass coasters. The woman from whom I purchased these thought the round glass bottom coasters were for eggs. She never could fit an egg into them, so the coasters were never used as intended. c. 1950s. $35-40.

Collecting - When I started collecting useful household linens in 1970, there were no books available to guide me through the pitfalls. My one saving grace was that I never bought damaged pieces. Once the fabric is gone, it is gone forever and a woven repair will always show. The exception, however, might be for an exquisite French linen and lace banquet tablecloth with twenty-four napkins sporting a cigarette burn but fairly priced — then I can live with the repair. If the piece fits your need, is the right size, and the perfect color, then collect it! At the same time, be selective: just because grandmother left you a purple and orange quilt that you know she bought at a church rummage sale over forty years ago and the colors don't quite match your decor, you're not required to add it to your collection. Give the quilt to another relative who likes it, or sell it and buy one that does fit your home.

A good reason for collecting vintage linens is that modern machine-made items are of questionable quality, while antique pieces have craftsmanship, quality, and design, yet are still affordable. By affordable, I am referring to useful, everyday household linens priced at $200.00 or less; there are thousands of such affordable, desirable items just waiting to be found. Look for well-made items that are beautiful both front and back, and of delicate design yet sturdy and strong. Edges should be finished off and not frayed or tattered.

Colored Damask - Colorfast dyes were used in the second quarter of the twentieth century to produce vast hues of colors, mostly for use on damask dinner tablecloths and matching napkins. Colored damask is very highly sought, thus commands a higher price than white damask.

Colored Damask

These three examples of colored damask are all of high quality linen. The light green lapkin is from a set of seven other lapkins and matching tablecloth, all with the "GWB" monogram. The tablecloth is 144"L x 72"W. c. 1920s. $150-175. *Courtesy of Whittier T. Brown and Barry L. Aldrich.*

A close up of the gold and white fringed 20" square napkin. c. 1890s. Set of eight, $40-48.

A lavender and white colored linen damask tablecloth
with a musical motif. 84"L x 68"W. c. 1920s. $85-95.

Compass Work - A type of eyelet embroidery consisting of overlapping circles that form a four-petal pattern. After running stitches were worked around the petal shapes, the openings were cut and buttonhole stitches were used. Often the center had an embroidered dot.

Condition - Linen, when used over a lifetime, takes on a certain sensuous feel, a soft patina that no other fiber can match. If you are fortunate enough to have acquired the family linen sheets, don't think of them as heavy, coarse, wrinkly old things. Once you sleep with linen sheets, you will never go back to anything else.

When purchasing vintage linens, put them up to the light. Use either a light bulb or direct sunlight to search for holes, thin areas, repairs, or overall wear. This is important if you want to be able to use your pieces. Do expect slubs, which are a group of thicker than usual threads and are characteristic of vintage-made linen (some looms were powered by water mills before electric powered mills were built, resulting in uneven threads). Due to the rise in demand, general line antiques dealers have now jumped on the proverbial bandwagon selling vintage linens, calling each and every piece Battenberg Lace. If you're a beginner, buy only from reputable dealers, as they will stand behind their merchandise and share their knowledge. Once you are more advanced and sure of what you're looking at, you can buy from general line dealers. Always beware of "here today, gone tomorrow" dealers at one day flea markets: you really do get what you pay for. The bottom line is that if the piece has missing lace inserts, damaged fabric, torn lace, many dark stains, bad repairs, and yellowed, torn creases, pass it by.

Cordonnet - An outline thread that defines the pattern of the lace, which can be either handsewn or machine made. Laces with cordonnets are Alençon and Carrickmacross.

Cotton - A seed fiber that in 1995 met fifty-four percent of the world's total fiber demand. Cotton, which grows on bushes 3' to 6' high, grows anywhere the season is long and the climate is temperate. This fiber is grown within a boll or seed pod. The blossoms appear, fall off, and the boll or seed pod begins to grow. Inside the boll are seven to eight seeds from which the fibers grow. When the boll is ripe, it splits open and the fluffy white fibers spread out.

A pair of linen compass work guest towels. Although compass work is not hard to do, I have been unable to find many other pieces. c. 1950s. $20-24.

Throughout the 1600s and 1700s, cotton fibers were separated by hand in the southern United States colonies. Then, in 1793, Eli Whitney invented the cotton gin. The gin could separate fifty pounds of cotton a day, compared to one pound a day by hand. The cotton is picked by machine, the fibers removed from the seed by a gin, and then pressed into bales ready for spinning. Cotton is absorbent, comfortable, durable, and has a pleasant texture. Highest quality long staple cotton, made in Pima, Arizona, is the source of today's fine percale cotton.

Counterpane - Also known as a summer quilt or summer coverlet, a counterpane is a quilted coverlet without any filling in between.

Coverlets - These are bed covers that cover only the top of the bed and need a dust ruffle or bed skirt. Counterpanes, quilts, Marseilles, and Matelassés all fall into the category of bed coverlets. Early beds were high off the floor to avoid cold drafts. High beds also stored trundle beds, which had wheels for portability and were usually used for children.

Crib Covers - These decorative yet useful bed coverings were used not only for baby's crib but also for the baby carriage. Most crib covers came with a matching pillow, again, mostly for decorative purposes.

Crochet - Known in Europe in the 1500s, this simple yet exquisite type of needlework requires only a hook and some thread. Crochet is one of the most common laces found in our mothers' and grandmothers' trousseaux.

Crib Covers

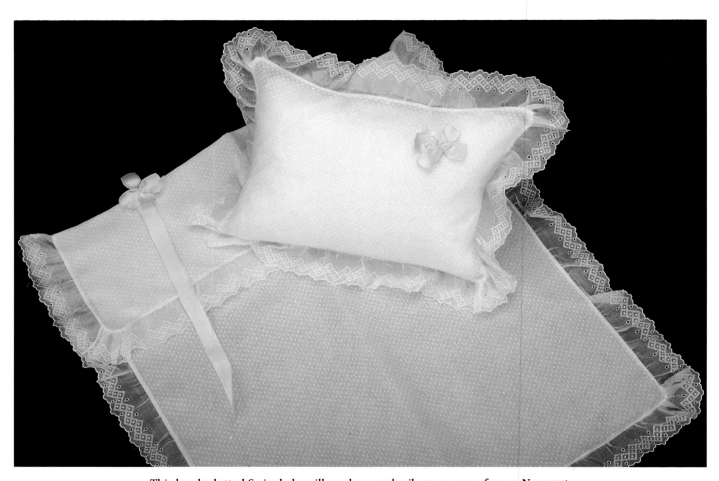

This lovely dotted Swiss baby pillow sham and crib cover came from a Newport, Rhode Island estate. Sets like these were very widely used for babies (as originally intended), but are now used for doll beds and carriages. c. 1950s. $40-50.

Opposite page bottom:
A cotton crocheted bedspread with two cut corners for a four post bed. c. 1930s. One of a matching pair of twin spreads. $175-200 for the pair. *Courtesy of Michael Mary Cowan.*

A filet crochet edged huck linen towel made in England. c. 1910. $15-18. Also shown
is a hand crocheted cotton thread runner in a very popular pattern. c. 1930s. $15-20.

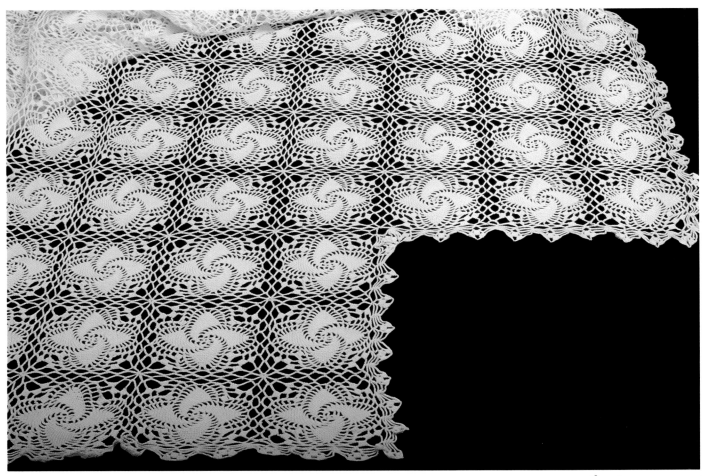

An ivory crocheted doily. c. 1940s. $8-12.

A crocheted doily in the pineapple pattern. c. 1940s. $8-12.

A white cotton crocheted doily in an Irish pattern.
c. 1930s. $8-12.

Crosstitch - This is a form of hand embroidery with each stitch forming an "x". Silk embroidery floss was most often used before the 1920s. After 1920, colorfast embroidery floss was used, meaning it could be bleached, boiled, and washed in hot water and the color wouldn't run or fade.

Crosstitch

Above:
Left: A pair of linen pillowcases with purple crosstitching in the pansy pattern. c. 1940s. $25-35. Middle: A home kit crosstitched linen table-cloth. c. 1950s. 84"L x 66"W. $25-35. Right: A nineteenth century home-spun linen pillowcase with the bride's monogram of "SMcK #8" (Sally McKenny's number 8 pillow-case, probably from a set of twelve to twenty cases). c. 1860. $45-55.

Left:
A queen size cotton top sheet proudly displaying the fine hand-work of its original owner's mono-gram. c. 1900. $75-85

A linen runner
made in China with
very fine colored
floss crosstitching.
c. 1950s. $10-15.

Curtains - Fancy window panels were used to filter sunlight, noise, visibility, and drafts, or just for the sake of vanity. During the Great Depression, my grandmother Burgess had very little furnishings in her home, yet everyone remarked on the loveliness of her lace curtains. Grandmother's curtains gave the impression that she had far more furnishings in her home than were really there.

Curtains

This is one of two pairs of machine-made netted background cream colored curtains. c. 1940s. Four panels $100-125.

A machine-made lace panel from England with a
fringed edge bottom. c. 1940s. The pair, $55-65.

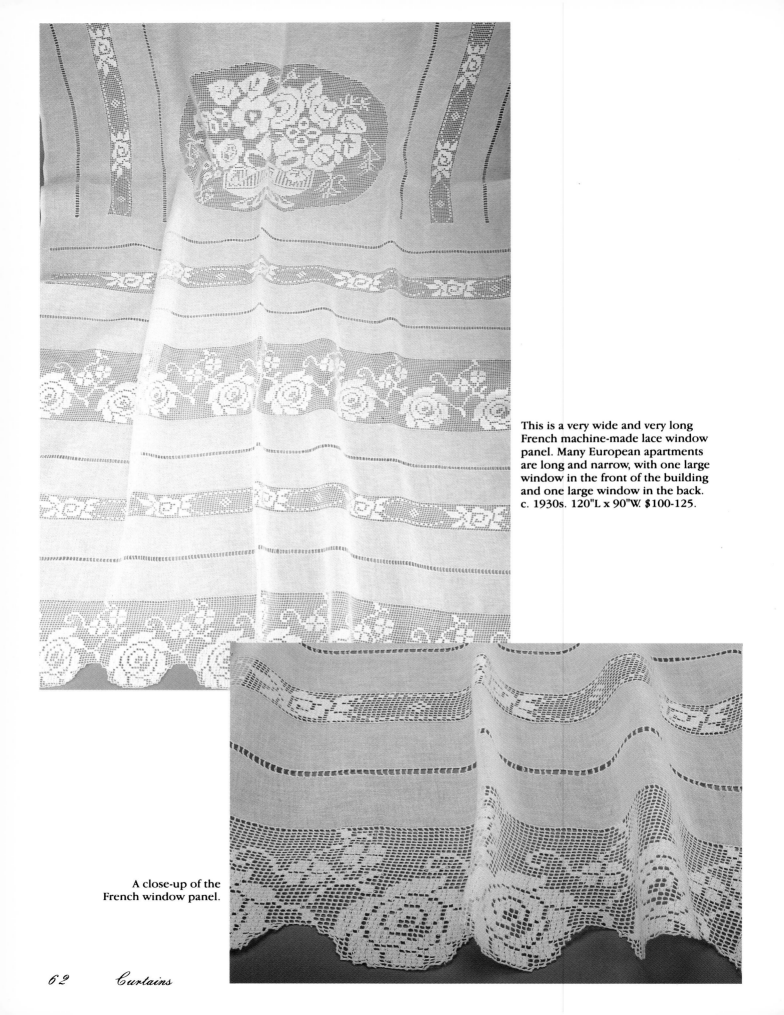

This is a very wide and very long French machine-made lace window panel. Many European apartments are long and narrow, with one large window in the front of the building and one large window in the back. c. 1930s. 120"L x 90"W. $100-125.

A close-up of the French window panel.

Cutwork - A technique of cutting out an area of fabric after a design has been outlined with the buttonhole stitch. Cutwork gained its highest popularity as a Portuguese cottage industry. Clever merchants would bring bolts of Irish linen im- printed with patterns to Madeira Island house- wives. After several weeks, the merchants would return to pick up the finished linens and to leave behind more bolts ready to be worked on.

Cutwork

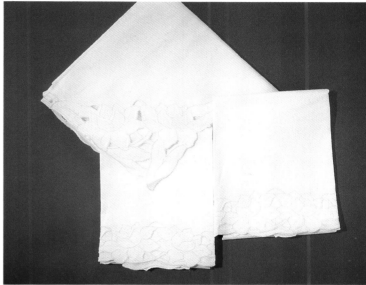

A cotton cutwork double sheet and two matching pillowcases. The cutwork pattern was an iron-on design from 1966. $45-55.

The top edge of this cotton sheet has all hand cutwork and embroidery in a paisley pattern. c. 1870s. $50-75.

This Italian cotton lay-over sham top has machine-made cutwork with connecting brides. c. 1940s. $40-50.

Damask - A fabric with a pattern that is created by using a weave different from that of the background, and that is reversible. Damask is woven on a jacquard loom. Like sheets, pillowcases, and towels, a fine damask tablecloth and napkin set was a household necessity.

Darned Lace - A knotted net background with running stitches forming a geometric design; another form of needle lace. Mass production of darned lace occurred largely in Italy around 1920. Once the Chinese recognized how inexpensively this popular lace could be made, they also mass produced large quantities meant for export to the United States.

Left: An Irish linen damask tablecloth and twelve matching napkins, hemstitched hems, in a fall harvest pattern. This set was never used and is still in its original box. Tablecloth is 108"L x 72"W; the napkins are 24" square. c. 1920s. $125-150. Top right: This roll of never used napkins measures 288"L x 24"W and shows what one dozen napkins look like before they are cut and hemmed. c. 1920s. $40-45. Bottom right: A gift box of one dozen fine Irish linen damask napkins. c. 1940. $50-60.

Darned Lace

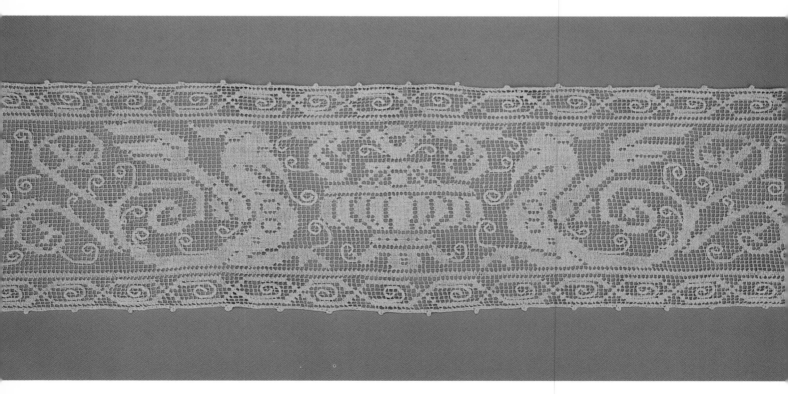

This ivory darned lace tablecloth and napkin set was made in China. The
geometric pattern looks lovely against a dark wood table. The napkins
are finely woven linen. Tablecloth is 90"L x 70"W. c. 1940s. $75-95.

Opposite page bottom:
An ivory cotton darned lace runner in the urn
and dragon pattern. Made in China. c. 1940s.
48"L x 17"W. $20-35. *Courtesy of Reneé Maury.*

Depression Era - Since few jobs were available for women in the Depression years of 1929-1939, women stayed home and used their time to crochet and knit tablecloths and bedspreads with whatever yarn, string, or twine could be found. Many beautiful pieces from this era are still stored in blanket chests and attics, not used due to their heavy weight. Full size crochet bedspreads alone can weigh up to 15 pounds!

Depression Era

Cotton crochet bedspreads were made by the thousands during the Depression Era. This spread has the popcorn stitch pattern in ivory cotton crochet. Whatever material was at hand was used during this era; beware of spreads made of nylon fishing twine! 110"L x 90"W. c. 1930s. $75-100.

Doily - A general term used for small decorative or utilitarian mats. They can be any shape, any size, and made from any fiber needed for a specific use. Doilies were named for a London textile merchant in the 1700s.

Doily

A machine-made net lace doily with a pretty pattern.
These were bought at 5¢ and 10¢ stores. c. 1950s. $10-15.

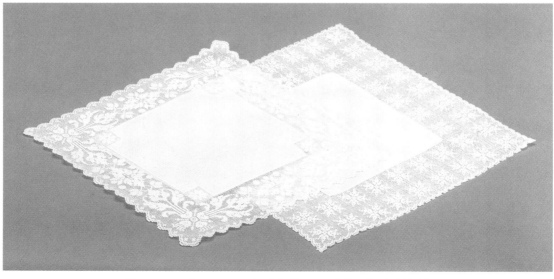

Two square Buratto worked doilies. c. 1920s. Each $12-15.

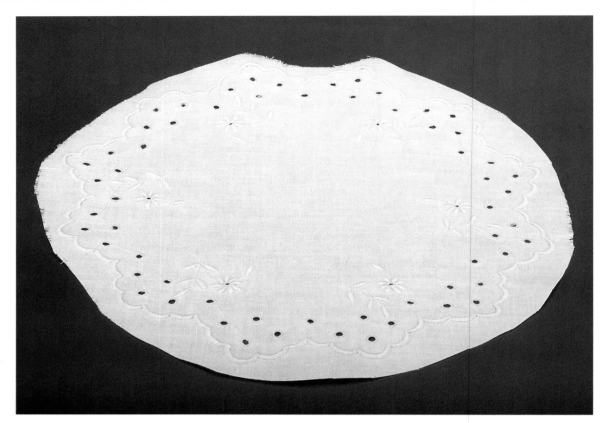

A simple embroidered, cutwork, and scallop edged round doily. The hard part was completed but the easy task of cutting out the doily was left unfinished. 20" round. c. 1940s. $10-15.

A round crochet doily with a very elaborate design. c. 1940s. $8-12.

Double Damask - The terms double damask and single damask refer to the looseness or tightness of the weave. Although both are single structure fabrics, double damask is the looser weave, with an overflow of seven threads, as compared to single damask, which has an overflow of four threads. Double damask, with 180 threads per inch thread count, produces a much finer and richer cloth. In comparison, single damask has 140 threads per inch thread count. Threads on double damask are also thinner than single damask, thus the cloth has a shorter life span than a single damask cloth. Despite the higher thread count, the weaving techniques used for double damask sometimes produce a low quality fabric, so the term double damask is no longer synonymous with "best quality." In reality, a good single damask can be superior in quality to a poorly woven double damask.

Drawnwork - In the twelfth century, a technique called drawnwork was produced by drawing counted threads out of the fabric in a certain area or by drawing threads together so that spaces were formed. With the remaining threads, fancy designs were created. Tenerife Lace, or wheel lace, is the handmade lace used in the corner of the drawnwork. Drawnwork is also known as Hamburg point, Dresden point, or Indian work.

Drawnwork

This linen drawnwork tablecloth was made from a kit. The hand embroidery is rather coarse and heavy but the overall effect works. 68" square. c. 1920s. $60-75.

Above:
This is a wonderful example of fine drawnwork that should be included in a collection. The fineness of the linen threads makes them easier to be drawn. Notice the Tenerife Lace work in the corners. 52" square. c. 1910s. $55-65.

Right:
A trio of neat little drawnwork doilies. These usually come in lots of three or more, all usually worked in different patterns (probably mini samplers of someone's work). c. 1920s. Each $8-10.

Dresser Scarfs - Bureau tops are often catchalls for nail polish bottles, perfume bottles, and loose change, so scarfs were made to protect the varnished finish or marble tops of bedroom case pieces. These scarfs often came in matching sets—a long scarf for the lady's bureau, a short scarf for the man's tall chest, and two smaller pieces for the bedside tables. Scarfs are also used as table runners and as window valances.

Dresser Scarfs

Although this dresser scarf has a machine-made lace border, the linen center is well made and considered very desirable. Notice how the corners were mitered ever so neatly to match the pattern. c. 1950s. $20-25.

A very tailored drawnwork
scarf on linen. 44"L x 17"W.
c. 1920s. $25-30.

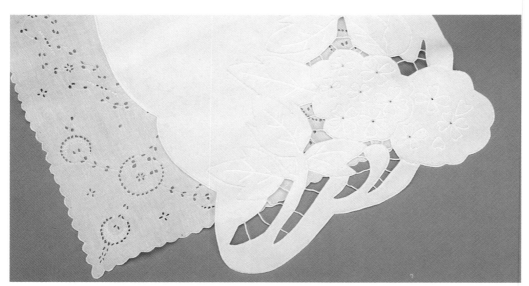

Left: This Madeira linen
dresser scarf has a very tiny
cutwork pattern that is worth
mentioning. c. 1920s. 36"L x
16"W. $15-20. Right: A linen
dresser scarf in the hydrangea
pattern, with brides connecting the open spaced cutwork.
This scarf has a pattern on
three sides only. c. 1950s. 42"L
x 18"W. $20-25.

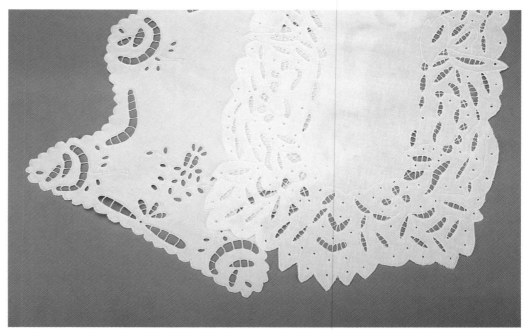

Left: An Italian cotton
cutwork scarf. The pattern is
very unimaginative yet the
workmanship is excellent. c.
1920s. 48"L x 16"W. $15-20.
Right: Another Italian cotton
cutwork scarf; this one made
by machine. Bought from
nuns in Lawrence, Massa-
chusetts. 44"L x 16"W. c.
1950s. $20-25.

Duchesse Lace - This "free" bobbin lace is from nineteenth century Belgium and consists of a noncontinuous thread technique made on a pillow. Bold patterns of clustered three-dimensional flowers and ribbed leaves work up quickly as a result of the heavy thread used for this lace. Individual motifs are made separately and then connected with bobbin-made braids. The wedding gown of Princess Grace of Monaco was fashioned of Brussels Duchesse type lace from the 1880s.

Duvet Covers - These are fancy European linen or cotton covers used to protect the feather down comforter or blanket. With some duvet covers, the comforter is slipped inside the duvet and buttoned closed on one end. Other duvet covers have a top piece with buttonholes only; the blanket or comforter then has the buttons. Unfortunately, most European duvet covers do not fit standard American blankets or comforters.

Duvet Covers

This duvet and pillow cover set is from Germany. The European pillow cover is 30" square and the duvet cover is 88"L x 38"W. This size just barely fits our standard twin bed if used as a sheet. The lace and fabric are cotton. c. 1940s. Two piece set $75-85.

Edwardian - Refers to the period from 1901-1910, during which King Edward VII reigned in England. He and his Queen, Alexandra, lived an ostentatious lifestyle, and linens of that time were lavishly embroidered with doves, laurel wreaths, ribbon bows, flower garlands, peacocks, good luck symbols, and hearts. Overall design and workmanship were of the highest quality.

Egypt - Prior to the turn of the twentieth century, the largest quantity of cotton worldwide was grown in Egypt; in addition, Egypt produced the highest quality of fine percale cotton at that time. Perfect climate and correct irrigation play an important part in the rich quality of good cotton fibers.

Embroidery - The ancient art of decorating fabric by using a needle and a large variety of fancy stitches with thread. Embroidery floss is the preferred thread and comes in a multitude of colors, but the most desired and sought after of vintage embroidered linens are those with white thread on white fabric, called, appropriately enough, white-on-white embroidery.

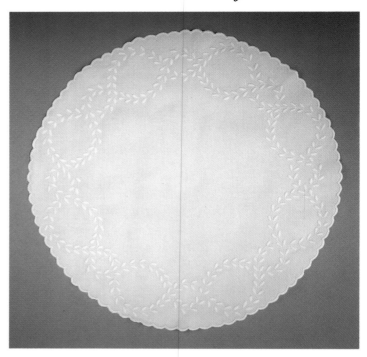

A doily embroidered simply with cotton thread on linen. c. 1920s. $10-15.

A very lovely French embroidered padded satin stitched three sided runner with a scalloped edge. c. 1920s. 46"L x 16"W. $30-35.

A hand embroidered guest towel with a whimsical winking owl. c. 1950s. 16"L x 12"W. $5-6.

English Lace - England's claim to fame came during the Industrial Revolution when its great invention of lace making machines took the textile world by storm. Even today, England's legacy of lace curtains still reigns.

Eyelet Embroidery - One form of English lace, this white-on-white machine-made lace design uses overcasting stitches around holes that have been punched or cut into the fabric. Also known as Hamburg Lace or Broderie Anglaise (English Embroidery). Extensive eyelet patterns were developed in England during the Industrial Revolution.

This is yardage of English Lace, also called Broderie Anglaise, eyelet embroidery, or Hamburg Lace. 244"L x 30"W. c. 1920s. $30-40.

Eyelet Embroidery

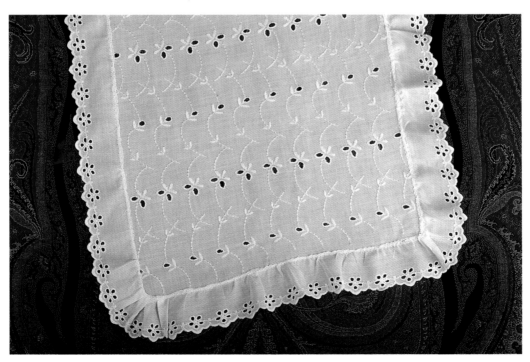

Eyelet embroidery has been manufactured for well over 100 years. This machine-made runner has cotton threads on cotton fabric. c. 1950s. 38"L x 16"W. $10-15.

Faggoting - After drawing out rows of weft (horizontal) threads, the sewer applies openwork embroidery stitches to connect two strips of fabric in a figure eight fashion. Also referred to as drawnwork, this technique was used mainly on towels and sheet sets.

Filet Crochet - Originally an imitation of net darning, this lace creates designs with open and solid squares. Filet crochet is a rather easy lace to make—and used for anything from dainty handkerchief trim to large bedspreads. Many fine examples of filet crochet come from England. English square tea cloths have beautiful lace trims of filet crochet.

Faggoting

This strip showing faggoting and a lace ruffle was meant to be added to the bottom of a pillowcase, petticoat, or apron. The drawn out threads are clearly visible, with the openwork embroidery stitches connecting the two strips. This piece is 36"L x 2"W. $15-20.

Filet Crochet

An English tea cloth with a machine embroidered fleur-de-lis design on linen and a diamond pattern filet crochet edge of handmade lace. c. 1915. 52" square. $60-75.

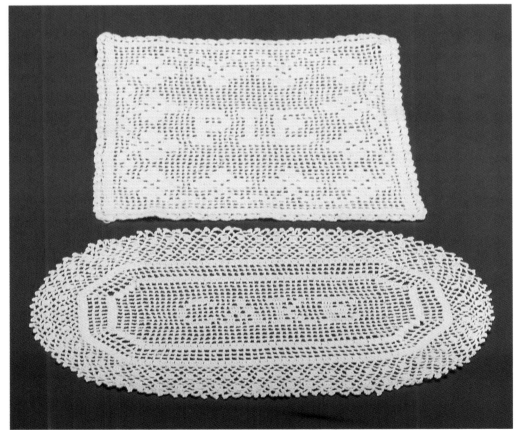

Above:
This is one of a pair of pillow-cases with extra long filet crochet lace. The pattern and the zigzag edge make this piece very desirable. c. 1920s. $35-45.

Left:
Top: A very unusual square filet crochet doily for a square pie. I have on occasion seen meat pies in square pans; this doily could also be used as a fly deterrent. 12" square. c. 1950s. $18-20. Bottom: This oval filet crochet cake doily was meant to be put on top of a loaf cake pan. The cake was then sprinkled with powdered sugar, or, as with the square doily, this one too could be used as a fly deterrent. 16"L x 10"W. c. 1930s. $18-20.

A filet crochet doily with an added touch of white swans. I had seen this pattern in books but never took the time to make one. When I finally decided to do so, I found this one in a box lot at a small country auction. I purchased it, so didn't need to make my own after all! 20" round x 4"H. c. 1930s. $50-60. *Collection of the author.*

A close-up of the swans with their
black beaded eyes and pink beaks.

Filet Lace - Although not a true lace, this style consists of an embroidery or darning stitch in linen on a network of square, knotted meshes, using a shuttle and a frame. It is actually much older than lace, having been known from antiquity. Also known as Rose Filet Lace, the netted background narrow-edged lace has a pretty pattern worked on the net. The edges are finished off in a buttonhole stitch for strength.

Fingertip Towels - When the morning's household chores were done, one often "received guests" in the afternoon to take tea, sew, or just socialize. Only the prettiest of guest fingertip towels would be put out when guests were coming for the afternoon. Hand and body towels were used when having overnight guests.

Filet Lace

An assortment of filet lace. The first, second, and fourth laces are to be used as edgings. The third lace is to be used as an insert lace. c. 1900-1950. $5 per yard.

Fingertip Towels

The pair of fingertip towels on the left are blue and peach, a popular bathroom color combination in the 1940s. The white towels in the center have light blue stitching for the embroidery. The pair on the right have organdy inserts and hand embroidery. All came from Madeira and are c. 1940-1950. Each set $15-20.

Flax - Flax is a prestigious fiber due to its limited production and relatively high cost, and was the exclusive domain of the wealthy. To produce linen from the flax plant requires art, science, craft, and a very special combination of both circumstances and highly skilled treatment. Since flax fibers extend into the root, the whole plant must be pulled by hand. After harvesting, the seeds are removed and made into linseed oil. To reach the fibers which lie under the outer bark, the stalks must be decomposed by water rotting. After the plants have been rinsed and dried, the bark is removed by a method called scutching. The fibers are then combed into a parallel fashion. The final stages include spinning, weaving, and, finally, finishing into linen. Russia is the main grower of flax but rarely ever exports any out of the country. Flax is also grown and produced in Ireland and western Europe. Linen's luxurious qualities are its body, strength, durability, and texture.

France - This country has an unsurpassed history of producing, weaving, monogramming, and embroidering some of the world's finest and most luxurious table and bed linens. The most knowledgeable and highly skilled textile workers have a family tradition of passing down their textile skills from one generation to the next. Some family members have worked for textile firms since the firm's inception; in some cases this dates back as far as the 1600s. The Porthault firm in Paris is still producing and creating one of the world's highest quality lines of linens to date.

Flax

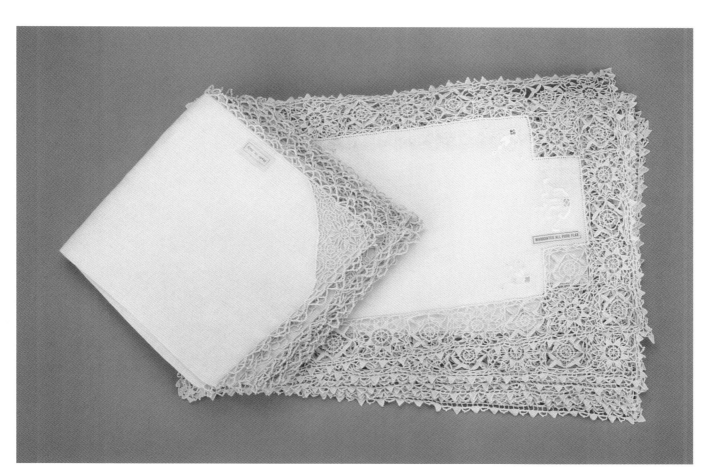

This is an elegant set of Italian Reticella Lace on natural, unbleached flax. These twelve placemats and twelve napkins have never been used. The paper label on the placemats indicates the linen is "Warrented Pure Flax." c. 1920s. $100-125.

French Embroidery - To do this "old favorite" very popular white-on-white embroidery, one must be skilled enough to make perfect eyelets and patient enough to make padded satin stitches precisely. These are the only two stitches required. The eyelets were first punched out and small buttonhole stitches sewn with a fine needle and thread. The satin stitches had to be sufficiently padded underneath for the final finish work on top.

French Embroidery

To make padded satin stitches so precisely is a highly skilled art. This is a classic example of perfectly made eyelets on white-on-white work. Table squares of this caliber should certainly be included in a collection. French, 66" square. c. 1910. $100-125.

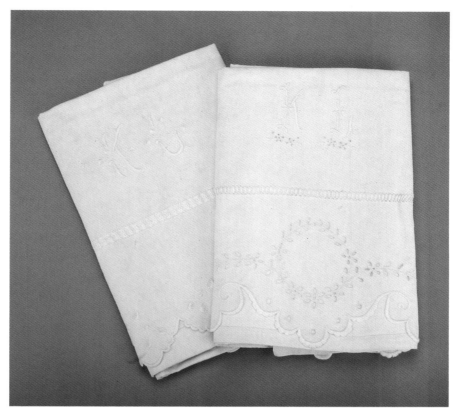

These are two twin sized cotton/linen blend top sheets with pretty cutwork designs on and around the padded satin stitched monograms. These were bought at a street fair in Châteauroux, France. c. 1940s. Each $40-50.

French Knot - This is just another one of many fancy stitches in embroidery. To form a French knot, one must come up from the back with the needle, wind the thread around the needle two to three times, then go back into the fabric, leaving behind a nice neat knot. When ironing French knots it is important to pad your board well and to iron the fabric on the *back* side to avoid flattening the perfectly made knots.

French Knot

A tailored white linen runner with two-tone blue flowers in French knots and green stems. c. 1920s. 48"L x 18"W. $18-22.

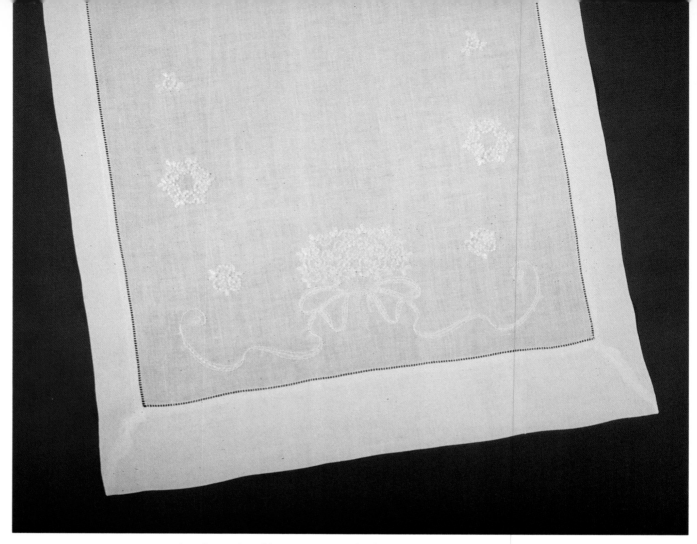

Another tailored linen runner with white French knots. The knots make a nice filler for the flowers. 48"L x 18"W. c. 1920. $18-22.

A close-up of the bouquet of flowers with French knots.

French Lace - Although the French "adopted" many lace designs from other lace making countries, they made their laces fashionably French. France had an influence in the making of Brussels, Mechlin, point de Venice, and Venetian Rose Point Lace. The most well-known laces from France are Alençon, Argentan, Chantilly, Lille, Arras, Normandy, and points de France Lace.

French Lace

Two ivory French net lace runners with rows of faggoting and appliqués of flowers. These are in demand today because the net mesh background is so fragile that few have survived. The one on the left has a linen center. c. 1920. 42"L x 17"W. $50-60. The one on the right is all net. c. 1920. 48"L x 17"W. $50-60.

This cushion cover is made of ivory Alençon Lace which is machine made. The ruffled edge is Paris Lace, also machine made. Note the original tag. 18" round. c. 1910. $40-50.

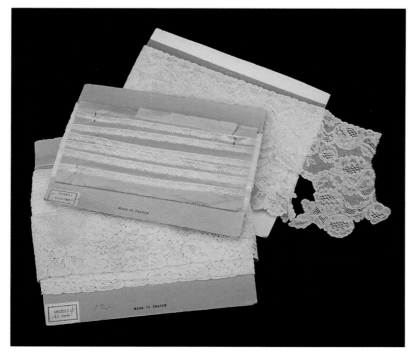

These are cards of French Lace made for the American market. Although the color is presently ivory, most will become white if washed. c. 1940s. $6-10 per yard.

German Lace - The first recorded example of handmade German Lace dates from 1526, when lace was brought into the country by Italian merchants from Venice. Then two books, a pattern book by Johann Siebmacher published in Nüremberg in 1597 and a book by Bernhart Jobin, published in Strasburg in 1583, illustrated the importance and the immense and lasting success lacemaking had at that period.

Interestingly, these pattern books rarely included technical instructions. Petit point and embroidery are still one of Germany's finest hand skills. "German Novelty" lace bed sets are machine-stitched tambour work. Examples of these ivory machine-made sets were common trousseau favorites during the late 1930s for young middle class working women desirous of the real thing.

German Lace

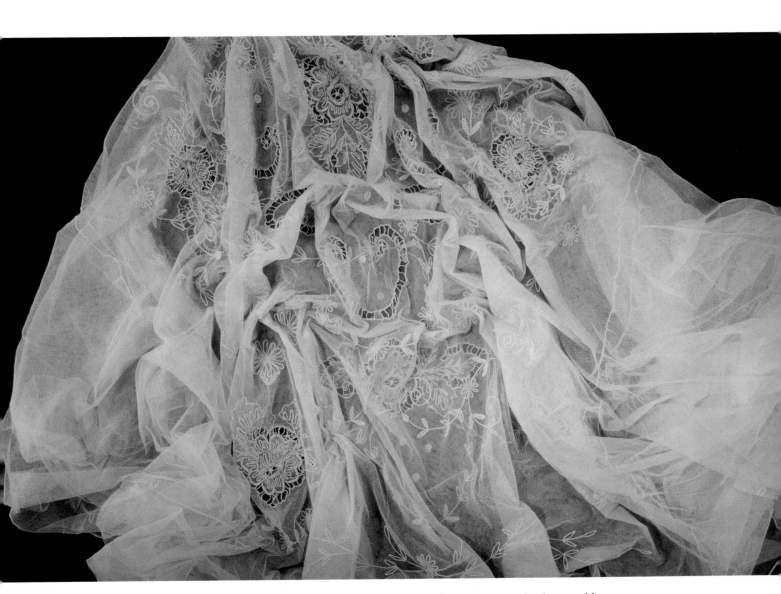

A double bed size German Novelty Lace bedspread. The ivory net background has machine-made tambour work. These sets often came with two pairs of curtains, bureau scarfs, bedside table doilies, and boudoir pillow covers. An underliner in a pastel color was often included in the set as well. c. 1930-1940. $150-200.

This is a small but wonderful example of German Petit point hand embroidery. The tailored linen handkerchief has a hand-rolled edge hem. 18" square. c. 1950s. $10-15.

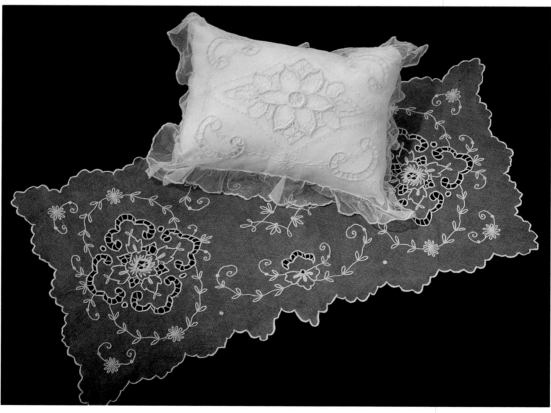

This boudoir pillow and bureau scarf are part of a German Novelty Lace bed set. Although the machine-made tambour work is very apparent, it makes a striking lace statement. c. 1930-1940. Boudoir pillow $40-50. Net Lace runner $30-40.

German Linens - Eiderdown covers and large 26" square pillow shams designed in Germany are an American favorite and highly sought. I have been told that people fleeing from their homes during bombing invasions took only their most precious items and had to leave everyday household items behind just to stay alive, thus very few vintage German linens exist today.

German Linens

The German pillow cover is 30" square with machine-made cotton lace on cotton fabric. The pillow is inserted from the back where the button closure is located. The eiderdown cover, or comforter cover, is a flat top piece with buttonholes on four sides. The eiderdown has the buttons attached to it which are then buttoned to the cover. Since our American standard comforters do not fit German eiderdown covers, we now use these as top sheets. c. 1930s. Pillow $40-50. Eiderdown cover $50-60.

Grenfell Industries, Labrador - The Grenfell Mission was active in Newfoundland, Labrador from 1900 through 1920. Dr. Wilfred Grenfell, founder of the mission, designed many textiles for the inhabitants to work on during their long winters away from fishing and trapping. The designs were of local scenes such as people, houses, icebergs, seascapes, and snowy backgrounds.

Grenfell Industries, Labrador

This piece has the "folksy" look characteristic of most Grenfell Industries' textiles. Notice the local people, the snow on the roof, and the ever shining moon. c. 1900-1920. 40"L x 18"W. $150-200. *Courtesy of Susan M. Traversy Antiques.*

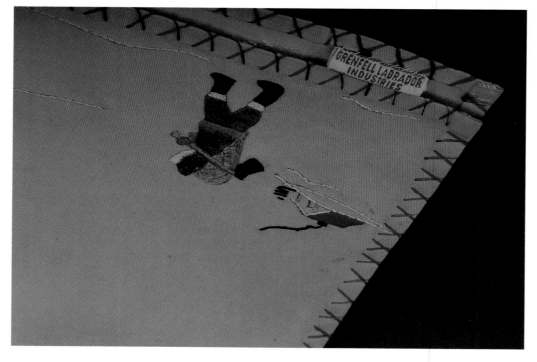

A close-up of the Grenfell label.

Hairpin Lace - This lace derives its name from the hairpin or fork-like instrument upon which it is worked. To make this lace, strips of loops are worked around parallel rods; special looms are also now available.

Hairpin Lace

Above:
This cotton top sheet with matching pillowcases has hairpin lace attached to the edges. The hairpin lace goes across the top of the sheet and down the side for 24" to make a nice fold over for the blanket. The long length of lace on the two pillowcases also adds to the prettiness for both the side and the top of the bed. The neat row of hemstitching adds to the tailored look of this set. c. 1940s. $60-75.

Right:
An interesting hairpin lace doily with a drawnwork linen center. c. 1920s. 12" round. $10-15.

Hamburg Lace - This is another name for machine-made eyelet embroidery lace. Although Hamburg, Germany, wasn't known for any particular handmade lace, the machinery capable of producing a form of eyelet embroidery lace was invented in this city. Made of cotton thread on cotton fabric, this white-on-white work became very popular and is still produced today worldwide, although synthetic threads and fabrics are used.

Hamburg Lace

Hamburg Lace was added to the edge of these percale cotton pillowcases. The almost sheer look of the Hamburg Lace makes a good contrast with the solid fabric of the percale. The pillowcases fit standard size pillows. c. 1940s. $25-30.

Opposite page top:
Left: A Madeira fingertip towel and hand towel with pink cotton appliqués. c. 1930s. $15-18. Right: A matching pair of ivory linen Madeira hand towels with cutwork. c. 1940s. $15-18.

Opposite page bottom:
Left top: An embroidered hand towel on huck linen from Madeira. Left middle: An embroidered and cutwork hand towel from Spain. Left bottom: An embroidered and Reticella Lace hand towel from Italy. Top and middle right: A pair of hand painted hand towels on huck linen from Ireland. Bottom right: An embroidered huck linen hand towel with Paris Lace from France. All c. 1950s. All $8-10.

Hand Towels - Proper household etiquette called for fingertip towels to be used in the guest bathroom for afternoon visiting guests and at the dinner table for washing fingers between courses. Hand towels, which were slightly larger than fingertip towels, were for guests staying overnight. Of course, only the finest were put out for their use.

Hand Towels

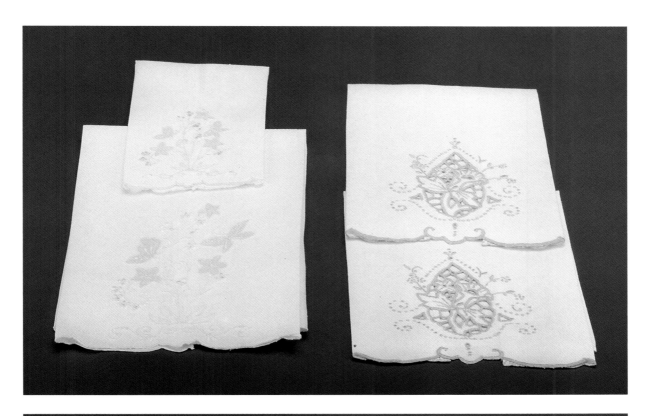

An Austrian huck linen hand towel with mosaic work and embroidery. c. 1950. $4-10.

A pair of linen hand towels with pastel embroidery, made from a kit. c. 1940s. $20-22.

Handkerchief - Long before the availability of disposable paper tissues, fabric handkerchiefs were the only means of catching a tear or dabbing a sniffle. To catch a gentleman's eye, a perfume scented hanky was purposely dropped near the gentleman's presence.

Handkerchief

The bobbin lace on this fine linen handkerchief is a Cluny Lace pattern. Notice how the pattern goes around the corner ever so neatly. c. 1900. 14" square. $20-25.

Handkerchief Box - These are triangular boxes for storage, usually made of glass and held together by satin ribbons so one could see through to find that very special hanky. They were very popular in the 1940s and 1950s.

Handkerchief Box

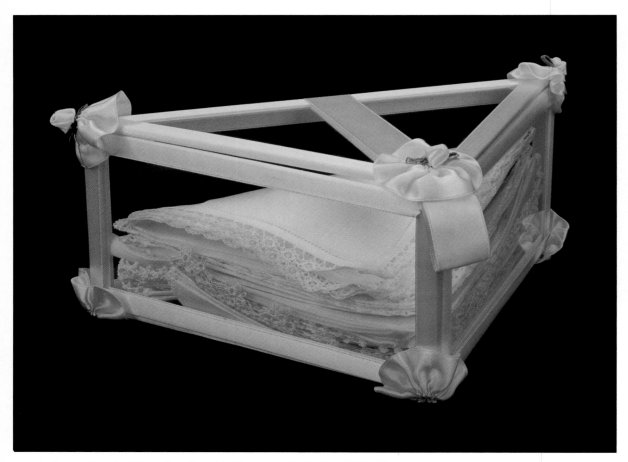

A glass and satin ribbon handkerchief box popular in the 1930s. These glass boxes made it easy to find that special handkerchief. $20-25.

A Flemish Art wooden handkerchief box with a lace edged handkerchief from Austria. c. 1920. Box $20-25. Handkerchief, c. 1950. $10-15.

A gentleman's aluminum handkerchief box, square shaped because men's handkerchiefs were folded in squares. c. 1940s. $15-20.

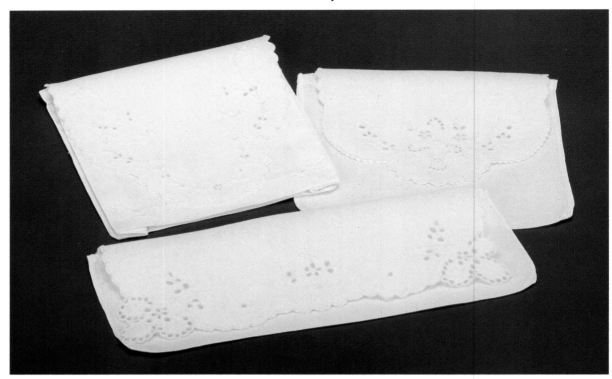

A trio of white linen Madeira embroidered cases. The top two are for handkerchiefs and veils while the longer case on the bottom is for gloves. c. 1915. Each $10-15.

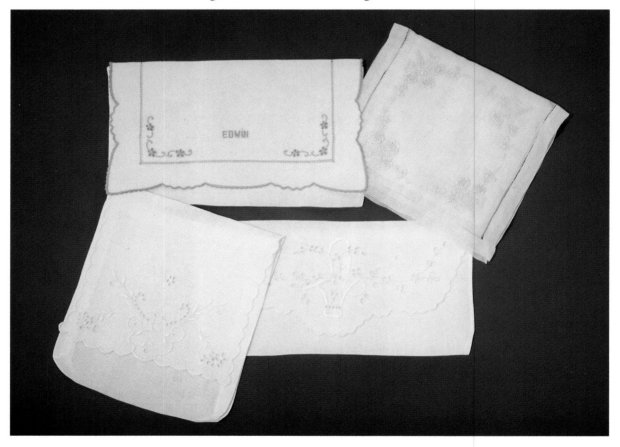

The Madeira embroidered case in ivory linen on the top left is obviously for a gentleman named Edwin. The top right case, made of organdy with a padded satin lining, has a satin ribbon tie to enclose the handkerchiefs securely. The case on the bottom left is for handkerchiefs, the one on the bottom right for gloves. c. 1915-1920. $10-15 each.

Handkerchief Case - Since ladies owned a large amount of handkerchiefs, square linen cases were used to store them properly. These cases could also be used for veils, an item also owned by every lady.

Hardanger - This counted thread drawnwork from Norway was worked with white thread on white linen and used extensively for household linens and traditional Norwegian folk costumes. The characteristic satin stitches are based on kloster blocks arranged to outline open spaces.

Hardanger

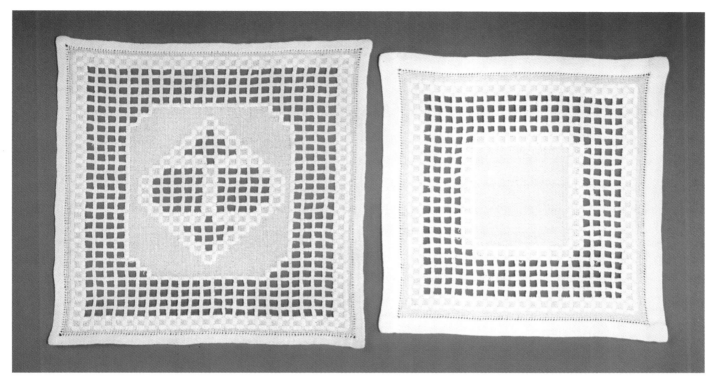

Above:
The squareness of the kloster blocks gives these Norwegian Hardanger doilies a distinctive tailored look. c. 1940s. $10-15 each.

Right:
A linen Hardanger runner with embroidered kloster blocks and drawnwork. c. 1930-1940. 46"L x 16"W. $20-25.

Heirloom - Refers to estate items passed down from one generation to another in the same family. One can acquire heirlooms from older relatives, whether living or deceased. Just because an item came from grandmother's attic or dowry, however, doesn't necessarily mean she made the item personally (unless, of course, you actually witnessed her doing so). Many times people mention that a specific item was made by their grandmother, yet upon closer inspection one finds an old store price tag attached! Not every woman was fortunate or skilled enough to learn each type of needlework known, and if she worked outside the home she was probably able to buy her ready-made linens at a store. Keep in mind, though, that even if grandmother didn't make a particular item by hand, it may still have its own history. A tablecloth, for example, may have been bought for a special bridal shower, a wedding reception, or a holiday dinner party.

Homespun - Before the invention of machine powered looms, fabrics were made on looms at home. The work was done either by oneself or by traveling itinerant weavers (who requested only room and board in exchange). Whether wool, cotton, or flax made into linen, the finished fabric was only 36" wide, the size of the common loom.

Homespun

This very interesting table mat is made from homespun linen. It has the finest embroidery designs made with silk floss, very intricate drawnwork, and a fringed edge. c. 1880. 36"L x 24"W. $25-30.

Honiton Lace - Made in the town of Honiton, England, this free bobbin lace has leaves and scrolls that are made separately then sewn together with fine thread. Some Honiton motifs are appliquéd to net or connected with needle lace stitches.

Hot Rolls - These unusually shaped pieces were placed in the bottom of the roll basket, the hot rolls put in next, and the four sides of the fabric brought up and around the rolls to keep the heat in. Similar embroidered pieces found are hot biscuits, toast, sandwiches, muffins, and bagels.

Hot Rolls

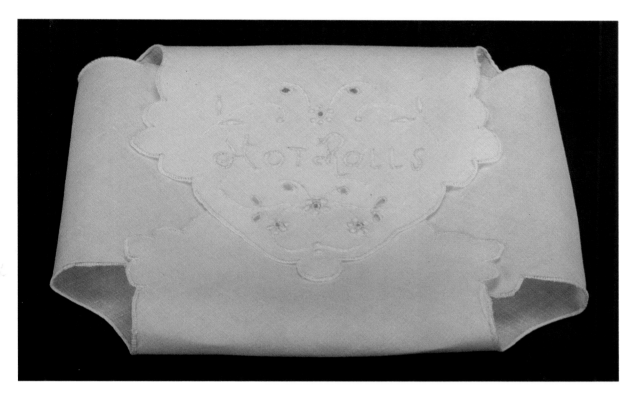

A linen hand embroidered hot roll cover, made in Madeira. Practical and useful, these covers are still manufactured today. c. 1920s. $8-10.

Additional covers designed to keep food warm are sandwiches, biscuits, toast, and muffins. The bagel cover would be unusual everywhere except perhaps New York City. Also most unusual is the baked potatoes cover with the embroidered potato. Each $15-20.

Huck Linen - Also called huckaback and made of medium to heavy Irish linen, this fabric is made on a dobby loom in a knubby, honeycomb texture for the best absorbency. Huck was used mostly for hand, body, and kitchen toweling. Toweling was available in great lengths and was cut to one's desired measurement.

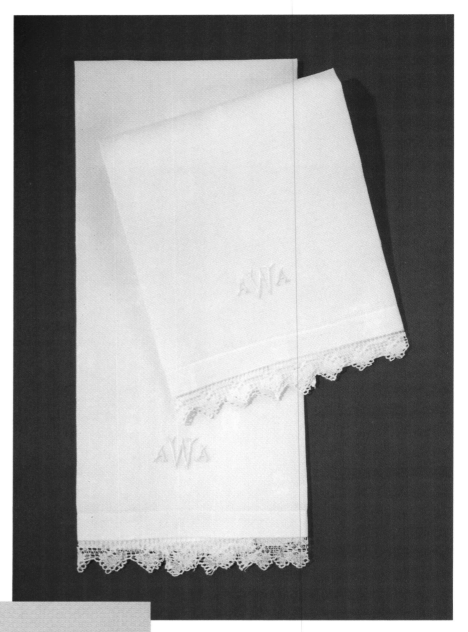

Above:
A pair of huck linen hand towels with a prestigious triple monogram and Rose Filet Lace for the edging. c. 1920s. $25-30.

Left:
A close-up of huck linen showing the nubby woven texture and three tailored monograms. c. 1920s. $10-15.

Illusion - An almost invisible machine-made net. Most bridal veils are made out of this tulle illusion, to give the appearance of a shimmering mist.

Imitation Lace - As the name suggests, this is an inexpensive and easy way of copying real lace. Imitation lace only came into being with the invention of lace making machines. Most pieces can fool the novice eye. There are some very fine examples available that are on net and are quite exquisite. A few that have been found with their original store labels on them came from the local 5¢ and 10¢ stores of the 1920s-1950s. The reason machine-made laces are called imitation lace, although they do have "lacy" designs, is because the lace-making machines don't manipulate threads like handmade lace. Machine-made lace, like handmade lace, is a decorative openwork of fabric with a design of spaces.

Imitation Lace

This machine-made lace runner was made with four different kinds of imitation lace. This is a good example of useful, practical lace that would be easy to care for in a vacation home by the sea or in the mountains. c. 1950s. 46"L x 15"W. $10-15.

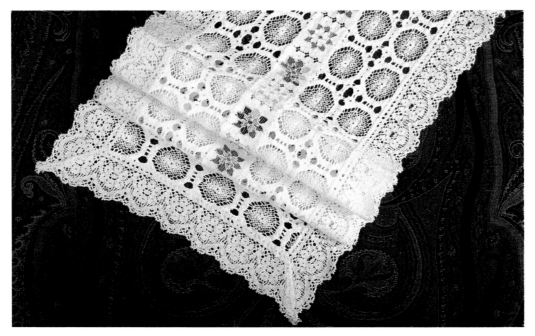

An ivory colored, machine-made lace runner of imitation lace. The band of color running down the middle dates this piece from the 1950s. 42"L x 16"W. $8-10.

This is a very fine
example of what
imitation lace
customers are
asking for. They
want the beauty of
real lace, without
the care or expense.
c. 1940s. 36"L x
16"W. $18-22.

Insertion Lace - Some laces are made right into the fabric but insertion lace is a swatch of lace made first and then inserted into fabric. To repair damaged quality linens today, pieces of vintage lace are placed over the hole and sewn on, then the damaged fabric back is cut away to show the newly inserted lace.

Ireland - When one speaks of the highest quality of linen produced, Ireland is the country that comes to mind (even though Russia actually produces the highest *quantity* of linen). Just as England is known for its fine woolens, France for its laces, and Italy for its handwork, Ireland is famous for its linen. Before technology was used to cultivate linen, those Irish fields of blue flax produced only one crop every seven years, hence the high cost, yet unsurpassed luxury, of this fabric.

Irish Lace - This is a very distinctive lace and has been copied, reproduced, duplicated, and imitated by every scheming lace merchant who isn't of Irish decent. The real version is a crochet lace with distinctive three dimensional raised roses, picots on the mesh, and a scalloped border. Irish crochet lace did not become widely known until the Irish potato famine of 1846, when children were sent home from school and men were sent home who worked the potato fields to make crochet lace for export in order to earn money for food.

Insertion Lace

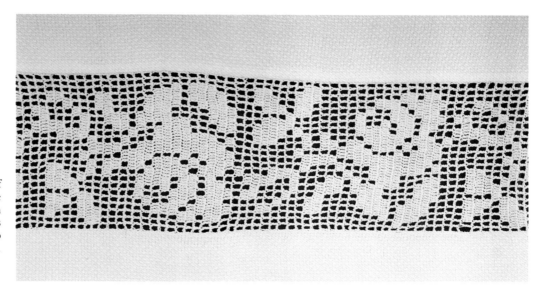

This example of insertion lace shows the Filet Crochet Lace with a rose pattern that was inserted between two pieces of huck linen.

Irish Lace

A pair of queen size Irish linen pillowcases with a row of drawnwork and a hand sewn border of white Irish Crochet Lace. The picots are evident, but the raised roses were not included. c. 1930s. $65-75.

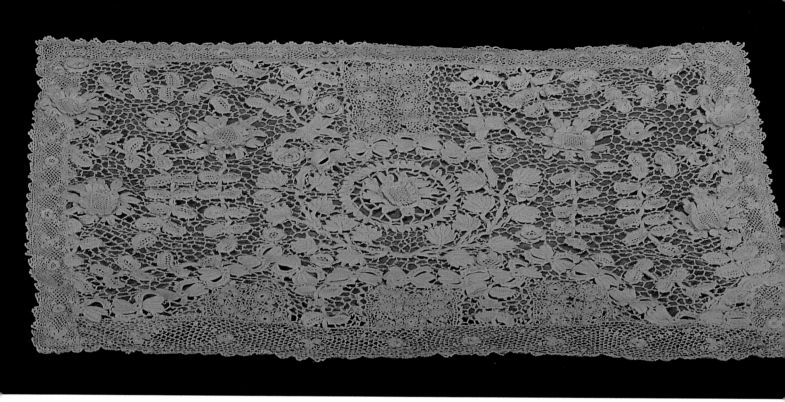

An Irish Lace pillow cover with distinctive raised roses, picots on the mesh, and a scalloped border. 24"L x 12"W. c. 1920s. $60-75.

Irish Linen - Linen manufacturers in Ireland can be traced to the thirteenth century. In 1685, Ireland's linen industry was given a substantial boost when exiled Huguenots from France settled in Ireland and incorporated their fine textile traditions into those of their newly adopted country. From that time on, Ireland has led the world in the production of beautiful linens. The skills and art of Irish craftsmen, whose lives have been dedicated to spinning flax, weaving fine linen fabrics, and bleaching and finishing the woven cloth, have been handed down from family to family. As a result, it is universally acknowledged that Irish linens have no equal.

Ironing - Just because ironing is a true labor of love for a select few of us, doesn't mean everyone else is going to enjoy the task! My method is truly quick and very gratifying if followed correctly:

First, the linens (everyday household linens like the ones photographed in this book, not those of museum quality) must be damp. After washing, make the decision to starch if the linens are to be used, or not to starch if they are to be put away. Place linens in the dryer on the very lowest heat setting for a very short time. Stand

at the dryer and keep touching them until they have just the slightest bit of dampness.

Next, use a dry iron. Anything embellished with cutwork or embroidery should be ironed on its reverse side on a thickly padded ironing board, not on your carpet or counter. Take your time. Enjoy the beautiful sight before your eyes. Your hands will be comforted by the warmth of the heat, your sense of touch will tingle at the feel of the linen's sensuous luxury, and your nose will be enchanted by the smell of the heat and moisture. I happen to use any iron on sale for $10.00, as long as it has 1200 watts (the wattage is on the footrest of the iron). Any scorching that occurs is not because your iron is too hot, but because there is leftover detergent in the fabric; a cool rinse of water will remove any resulting discoloration. When ironing lace or net, use a press cloth so the point of your iron doesn't catch the delicate work.

When your work is done, you will be rewarded with a stack of crisp, freshly ironed vintage linen, the very reason you so passionately seek out these hidden treasures tucked away in an old attic or in a box under a table at a flea market.

This set of napkins and matching tablecloth was made by the Old Bleach Company in Ireland. The colored flowers are hand painted and are guaranteed not to fade from either washing or light. c. 1940s. Napkins are 22" square, tablecloth is 108"L x 70"W. The set $100.

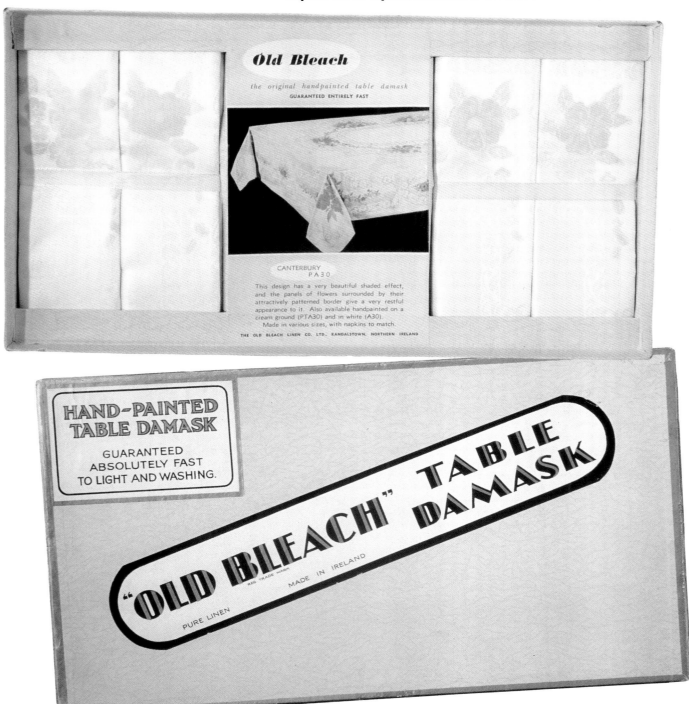

The front cover of the box in which the hand painted napkins and tablecloth were packed.

This wonderful set of double damask Irish linen dinner napkins still retains its paper label and satin ribbons. The one dozen napkins are 22" square. c. 1940s. $75-100.

Italian Lace - Venetian embroiderers of 1540 invented needle lace, and nuns from convents in Italy are world renown for their exquisite refinement of the lace-making process. The Italian nuns are also known for teaching this skill to homeless children in order to help them become craftsmen as adults. Bobbin Lace was almost certainly invented in Venice at the end of the first quarter of the sixteenth century. Exactly when and where the transition from one technique to the other took place is not known.

Italy - During the "gilded age" of the 1920s it was fashionable to travel abroad for a leisurely two to three-month stay. During this time, enterprising Italian merchants employed homemakers, as part of a cottage industry, to make linens for these tourists to buy and bring home. Italian linen made specifically for the tourist trade has a "Made in Italy" stamp or paper label to distinguish it from linens made for the Italian household.

An Italian 88" round linen tablecloth with Filet Lace insertions of flower baskets. The hand embroidery is of average workmanship while the cutwork is rather finely done. The border has figural Filet Lace. c. 1920s. $500-600.

This is one of a pair of Italian pillow shams with the comforting word *"Felicita,"* meaning peaceable. Items such as this were made in rural areas rather than in cities, where fancier was better. The padded satin stitching and the many different types of embroidery make this simple pillow sham a rather lovely piece after all. c. 1930s. $50-75.

An Italian doily, 36" round, with cutwork, embroidery, needlelace, and a figural design. Linens with figures command a higher price due to the recent surge of interest in collecting such hard to find pieces in fair to good condition. c. 1910. $50-60.

This pair of Italian pillow shams has very intricate red, white, and blue embroidery. The maker's monogram is "MP", for Mary Panelli of Bologna, Italy. c. 1919. $75-100. *Courtesy of Susan Curran McCahon.*

Made in Italy for the
tourist trade, this
queen size sheet has
a center medallion
with an embroidered
bow, cutwork,
needlelace, and
embroidery. Cotton.
c. 1920s. $65-85.

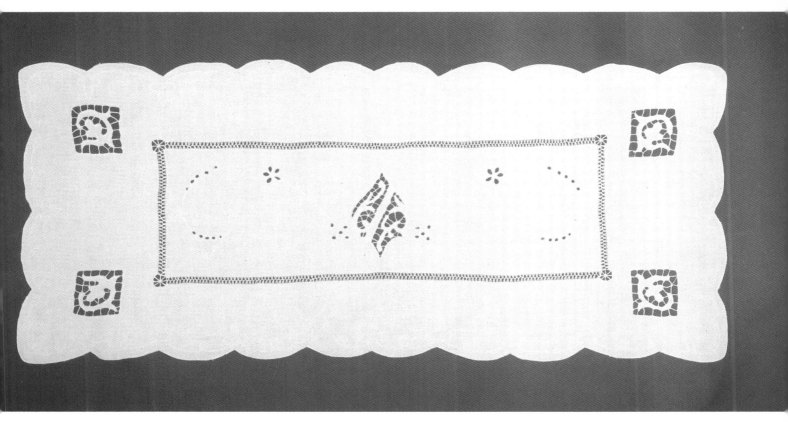

This matching runner to the queen size Italian cutwork top sheet
shows the same workmanship. 48"L x 18"W. c. 1920s. $20-25.

Jacquard Loom - Fabrics made on a jacquard loom include damask, brocade, and tapestry. Damask is the flattest looking but is unsurpassed in quality and durability due to its high fiber content.

Jacquard Loom

Made on a jacquard loom, this Early American linsey-woolsey coverlet is dated 1830 and was made by "H.S." The ivory colored linen side is used in the summer and the indigo blue wool side is used in the winter. 90"L x 80"W. $750-850. *Courtesy of Mimi Duphily.*

Japan

Japan - To get on the bandwagon in the 1940s and 1950s, Japan began exporting a rather low quality line of bridge sets, sheet sets, and rayon tablecloth sets. The only specimens left of these linens are ones that have never been used. I would advise not purchasing these as the low quality and poor workmanship do not stand the test of time.

Kitchen Towels - Since naturally grown flax is spun into long linen fibers, linen kitchen towels - a household staple - are lint-free and perfect for wiping glasses, tableware, and dishes. Some linen towels are manufactured with the word "glasses" printed on them and are meant only for glassware.

This pair of end table doilies with tambour machine stitching on cotton was made in Japan around 1950. When I purchased them, they had never been used or washed. After just a gentle soaking, however, the tambour stitching tore away from the fabric. There is no value to these.

Kitchen Towels

Here is an assortment of linen kitchen towels. The bottom towel has some crosstitching and is meant to be used as a tea towel. All are made of linen. c. 1900-1950. Each $5-6.

Kits - For those women around the turn of the twentieth century who were not fortunate enough to be taught the gentle art of handwork, home kits were available from ladies' magazines. Battenberg Lace, for example, was one type of design available in a home kit. The cotton fabric and machine-made tape lace were enclosed in the kit along with detailed instructions on completing the spider work.

Kits

This is a pattern on pink muslin for a doily kit that needs twelve yards of tape to make the outline pattern. c. 1915. $10-15.

The beginnings of a tape lace doily on pink muslin made from a kit. The tapes have been basted down on the muslin, and some of the lace fillings have been started but were never finished. c. 1915. $10-15.

Above:
Another pattern, this one for the use of silk embroidery floss. It was started but never finished. Notice the clever advertising on the corners. c. 1900. $15-20.

Left:
Close-up of the corner advertising.

This is a four piece tableware set that has a felt lining to protect silverware from tarnish. This set came from a prestamped kit. c. 1930s. $35-40.

Another prestamped kit for tableware.

Knit Lace - When knitting is done using a fine thread, thin needles, and with a continuous thread made by pulling one loop of thread through another, the result is lace. All the stitches are held together as loops on two needles. With access to the finest yarns and needles, knitting has developed into a highly skilled art form. Silk, cotton, and wool yarn are all used in knitting. Wool is the preferred fiber, as warm clothing was primarily made before lace. Knitted lace was never as fashionable as crocheted lace for household linens. As the popularity of knitting has revived in the last decade, several knitting guilds have recently been formed.

Knit Lace

A linen and knit lace runner with tassels bought at a Paris street fair. The knit lace is extraordinary due to the number of patterns and designs incorporated into the center medallion and border. c. 1920s. $50-60.

Knotted Lace - Originally called Armenian Lace, this lace comes from Armenia at the eastern end of the Mediterranean Sea. Girls at a very young age were taught the art of knotted lace for clothing and household linen decoration. The lace is worked with a sewing needle and requires the skill of twisting the thread around the needle. No gauge is needed and only one knot is used, quite similar to the knot mesh of netting.

Know Your Textiles - There are a myriad of ways for beginning collectors to increase their knowledge and understanding of textiles. Many museums have inventories of laces and linens and are starting to exhibit them. Craft supply stores in some areas are starting to hold handiwork workshops. Search out the many Victorian theme magazines available at the newsstand: they are full of great reading material and inviting photos, and contain lists in the back of merchants who deal in all sorts of textiles. While shopping at antiques shows, pick up the monthly antiques newspapers available at the front counter. They have interesting articles and list upcoming auctions and other antiques shows. Scan "for sale" classifieds in your local newspaper for items that may relate to your collection. Lace and linen societies do exist, and the members are eager to share their expertise. Names of such societies can be found in periodicals from the craft shops and trade journals. Last, but not least, know your established linen dealer. In general, flea market dealers are "here today, gone tomorrow" and cannot stand behind their merchandise if there is a problem with size or condition. You may, however, find linen dealers at flea markets who have business cards and appear reputable; use your judgment in such cases regarding whether or not to trust the dealer.

Knotted Lace

The Knotted Lace on the left is a 48" round doily with a hole in the center so large the doily can also be used as a parasol cover. c. 1900. $40-45. The Knotted Lace on the right is a camisole top to which the fabric body was never added. Notice the delicate but sturdy nature of this linen threaded lace. c. 1910s. $15-20.

Lace - Simply said, lace is an openwork fabric with yarns twisted around each other to form a pattern. Flax, cotton, silk, gold, silver, or mohair can be looped, plaited, or twisted with needles, bobbins, or by machine. The origin of handmade lace is obscure. Flanders and Italy both claim its invention, although Italy was the first to publish a pattern book. No other fabric, not even silk or velvet, has carried through the centuries the evocative power of lace. It is a fashion accessory and a status symbol, with a price that surpasses that of precious jewels. Lace has even been the subject of laws prohibiting its wearing by certain classes of society.

Lace

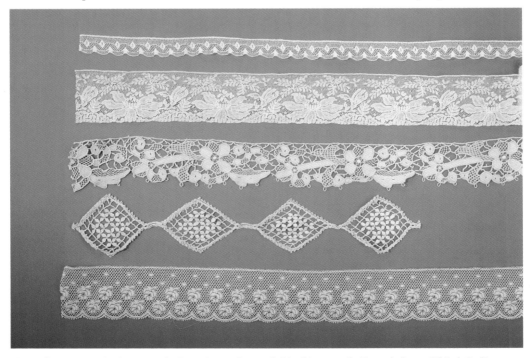

Top and middle: Knit lace. Bottom: Irish Crochet Lace. All $10-15 per yard.

First, from top: Machine-made Paris Lace. Second: Machine-made French Lace. Third: Italian needlelace. Fourth: English silk Maltese Lace. Fifth: Machine-made Paris Lace. All $10-15 per yard.

Insertion of *point de gaze* needle lace in Belgium Duchesse
bobbin lace. The petals of the flowers were made separately and
sewn on top of the mesh background to give added dimension.

An imitation lace collar made by machine
to resemble real handmade lace.

A fine example of needle lace made in Belgium. Often called "modern" lace, such lace dates from the period 1880-1910.

A woman's lappet of Flemish design in Russian Lace.

A flounce in ivory bobbin lace.

Lambrequin - A one edge or three edged valance or cornice used as a window sill, mantle, or shelf cover. Typically found in more than one piece, lambrequins usually will not fit your specific need.

Lambrequin

A French moiré lambrequin with silk ribbon embroidered flowers and a metallic braid trim. Used as a window valance or a mantle drape. c. 1900. $200-300.

Lapkin - A napkin measuring over 26" square that is large enough to completely cover your lap, used especially for buffet dining where you eat from a plate on your lap rather than sitting at a table. Another form of a lapkin is a napkin that is longer than it is wide, or is rectangular in shape.

Linen - The word linen is a generic term for all household textiles no matter what fabric is used. In addition, fabrics made from the flax plant only are truly linen. In its utilitarian capacity, linen fabric is prized for its strength, durability, and resilience. After long and loving use, it acquires a silken sheen and a fine patina that evokes refinement and comfort. Once reserved only for royalty, this now affordable domestic fiber is more readily available for everyday household use and clothing.

Lingerie Bags - These envelope type bags that we now use as boudoir pillows were originally meant to have pajamas placed inside them during the day to make a pillow. The satin or silk night clothes often worn today are not appropriate to fill up a lingerie bag, so pillow forms in feather or fiberfill are used instead to make a smooth-looking pillow.

Luncheon Cloths - Most often square in size, these fancy lace embellished cloths were meant to be used for guests rather than everyday household use. Of course, one's finest cloths with matching napkins were always used for these special occasions.

Lapkin

An ivory linen lapkin with a beautiful triple monogram and bobbin made lace trim. 26"L x 14"W. Set of six $75-85. *Courtesy of Rose Ewas.*

Lingerie Bag

An English lingerie bag made of organdy with Bedfordshire Lace edging and tambour work poppies. There is a pillow form inside the bag since it is being used as a boudoir pillow rather than as a bag for lingerie. c. 1930s. $40-50.

Luncheon Cloths

A dainty 38" square luncheon cloth made in China. The cotton cloth has appliquéd flowers and hand embroidery. c. 1950s. $25-35.

Machine Chain Stitches - Made as a basic chain stitch, machine chain stitching can be produced on net or fabric. Sometimes known as Swiss Tambour, it is also considered an embroidery stitch. Many linens with machine chain stitching are still being produced in Switzerland.

Machine-Made Lace - With the invention of machine-made lace, lace became available to all classes of people; no longer was lace considered a wealthy status symbol. As the production of machine-made lace grew higher, handmade lace fell to the wayside. With the machines, it was now possible to produce many new innovative designs in much wider widths.

Machine - Made Lace

A machine-made lace runner with a net background. If you want the look but not the price of real lace, this is for you. The ivory color provides the perfect antique "look," and the flowing, curvaceous design is most desirable. 48"L x 16"W. c. 1940s. $30-40.

These net background runners with a needle run lace design are machine made yet evoke an elegance that could grace any antique dining room table or bedroom dresser. Again, elegance without the cost. American made. c. 1930s. $25-30.

This machine-made lace imitates fine Italian handmade lace. A quality piece despite its lack of handmade origin, the figures and quality linen make it a desirable addition to any collection. 68" round. c. 1920s. $100-125.

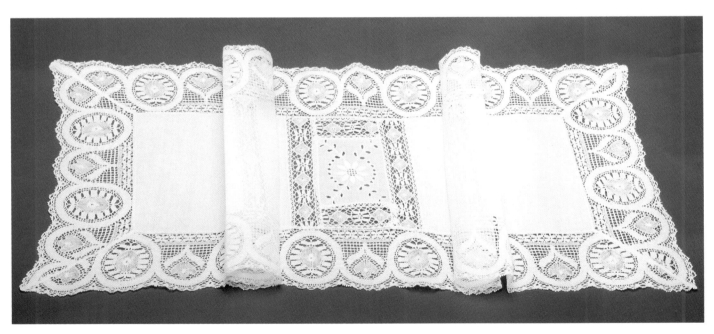

An inexpensively made cotton runner that has two types of machine-made lace. Purchased in 1958 at a dry goods store in New Egypt, New Jersey, for $1.19, this runner was bought by my godmother as a gift for my 10th birthday. American made. 48"L x 17"W. Value to me is priceless, but to you $20-25.

Macramé - Usually made from heavy or coarse threads, macramé is knotted lace made by hand; no tools are required. Most fringe seen on towels has a form of macramé as the lace work.

Madeira - The Portuguese island of Madeira, off the coast of Morocco, is world-famous for its superb skill of white-on-white embroidery and cutwork. Around the turn of the twentieth century, enterpris-ing merchants started a cottage industry on this sleepy little fishing island. Armed with bolts of white Irish linen, they would drop off the fabric to island bound housewives to skillfully embellish, then return in a month to pick up the finished linens. Most handwork was white embroidery on white linen, but examples of light blue embroidery thread on gray pencil sten-cil can also be found.

Macramé

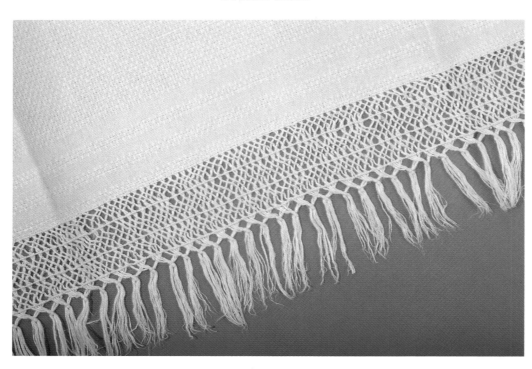

This is an edge from a set of three Irish huck linen towels. The macramé work is knotted and fringed by hand. c. 1900. $20-25.

Madeira

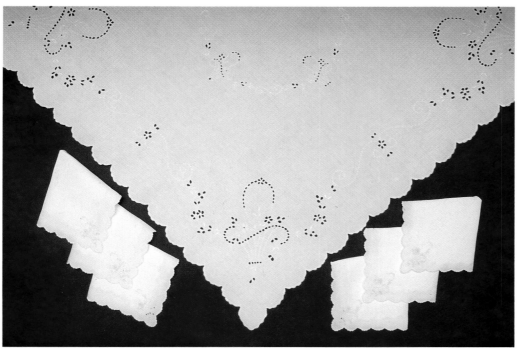

A white linen Madeira tea or bridge cloth with six matching napkins. The very popular basket motif is cut and embroidered in white thread. 48" square. c. 1920s. $25-35.

This white organdy tea cloth was made in Madeira. Notice the appliqués and shadowwork, as well as the very fine hand embroidery. It's very desirable to acquire tea cloths with matching napkins in sets of six. c. 1950s. $35-45.

This beautiful hand-embroidered queen size top sheet and its four matching pillowcases have the grey pencil stencil outline underneath the white thread of the embroidery. After repeated washings the pencil marks will eventually fade. This is a wedding sheet set from 1940 that was never used. $75-85.

This Madeira top sheet and two pillowcases, hand-embroidered with light blue thread, was a wedding sheet set from 1942. The set was used for one week only during the couple's honeymoon at a cottage on Cape Cod. The husband never returned from war, but he left his wife a legacy of twin sons, and she never remarried. Fifty years later, she sold the sheet set. $75-85.

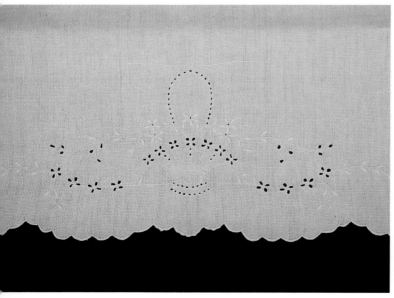

The needle workers on Madeira Island favored the ever popular basket design and used it on many household linens. Today, this design is just as much in demand as it was back then. 30"L x 18"W. c. 1920-1950. $10-15.

These two bathroom towels were sold as a set in the 1940s, when mixing and matching colored linen was all the rage. Set $15-18.

Maltese Lace - Created in Bedfordshire, England, this bobbin lace is made of silk and has motifs of the Maltese cross as well as a large number of *point d'esprit* (leaves).

Maltese Lace

Above:
An ivory silk English Maltese ladies lappet of bobbin lace, worn around the neck. c. 1900. $50-60.

Right:
A collar made of white silk English Maltese Lace. The crosses resemble the Maltese cross, and this resemblance (rather than any connection with Malta) led to the name of this lace. c. 1900. $50-60.

Mangle - An electric ironing machine with a heated roller to press linens. During the period when able, silent household staff cared for linens, the mangle was primarily used for sheets and large table-cloths.

Matelassé - A modern version of the vintage Marseille coverlet. Named for the French word meaning "padded", it is a machine-made white-on-white coverlet made of double cloth, as if weaving two fabrics of the same weight in a single operation. The coverlet has a quilted appearance that is created by puckering, which is produced by using patterning threads that are pulled tight. It is a hard-wearing cloth with an embossed appearance, formed by adding an extra weft in coarser thread. This causes the finer yarns on the surface to pucker, giving the quilted effect.

Moiré - This fabric is used for window, mantle, and table treatments. Made of taffeta, it has a watermark or a wood grain texture.

Moiré

The wood grain or watermark pattern on this moiré lambrequin provides a special effect on the silk ribbon embroidery. c. 1900. $200-300.

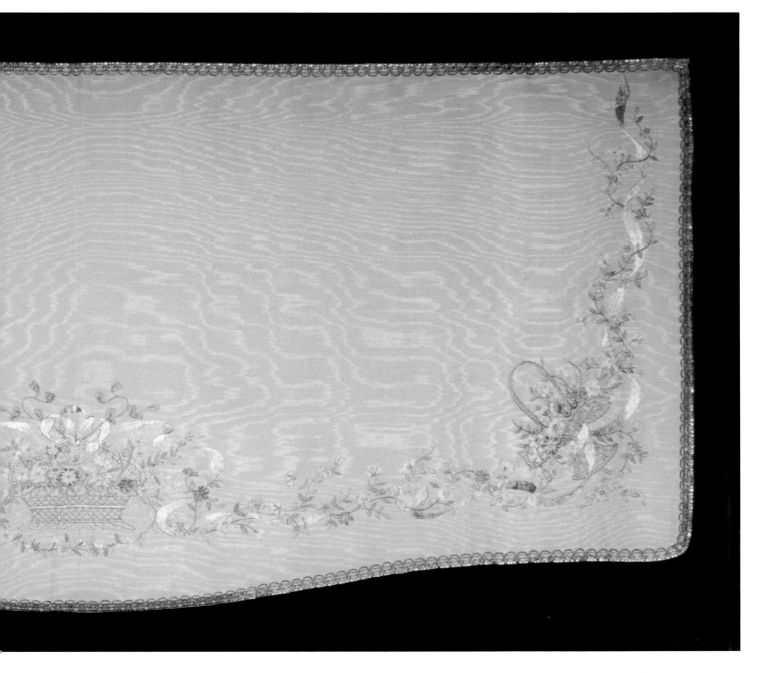

Monograms - In nineteenth century Europe, only royalty and nobility were fortunate enough to have their initials or coat-of-arms hand embroidered by the deft fingers of linen maidens. By the end of the nineteenth century, people realized that a woman's worth was determined by the large amount of fancy monogrammed linens she owned; she was now eligible to snare a willing suitor into marrying her. When assembling her personal trousseau, the young maiden, having been taught the art of embroidery by her mother, would place her future husband's initial before her already embroidered two initials. Monograms guarantee that the name of the forebearer will not be forgotten by descendants. Today, members of the aspiring middle class not only have their linens monogrammed, but also china, crystal, and silver.

Monograms

Above:
These four large bath towels, used before Turkish toweling was invented in the 1920s, are made of huck linen and proudly embellished with the original owner's monogram. Each $15-20.

Left:
This hand towel of huck linen sports a monogram of EST in the center of a cutwork wreath. c. 1940. $12-15.

A pair of twin size cotton German bed sheets with machine-made lace; the monogram of HR is done with padded satin stitches. c. 1940s. $65-85.

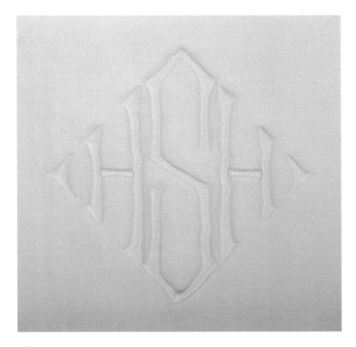

Above:
This bold, tailored monogram belonged to Hattie Smith Hall, Smith being her maiden name, Hall her married name. c. 1929.

Right:
This delicate hand towel with Paris Lace has a net filet insert with the monogram of EMV. c. 1900. $15-20.

Mosaic - Also called punchwork, this open spaced work was created by using an awl or stiletto to separate the threads and make an opening, then stitching around every hole using an overcast or a buttonhole stitch. I once purchased a trunk full of this fine handwork, with most pieces having original paper labels bearing the words "Made in Austria" as well as their original price tags. The original owner bought them on a trip to Europe in 1923. Although mainly called Austrian punchwork, many fine examples of mosaic were also made in Italy for both the tourist trade and for export to the finer department stores in America.

Mosaic

From a set of eight placemats, this mat starts out with a scalloped edge, moves to an area of fine mosaic work, then an area of larger mosaic work, and finally a row of tight embroidery, all on linen. 16"L x 12"W. c. 1920s. Set of eight placemats and napkins $60-75.

A corner of a very large tablecloth that has many mosaic holes, some embroidery, and just a little cutwork. The numerous mosaic work holes make this a truly fine example of craftsmanship. 128"L x 70"W. c. 1930s. $125-150.

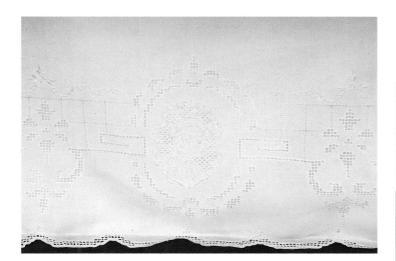

This is a close-up of a boudoir pillow with two silhouettes. The mosaic work and the silhouette embroidery nicely complement each other. c. 1920s. $35-45.

This corner is from the other half of the boudoir pillow.

Napkin - A common everyday necessity, napkins are used completely open and on your lap to catch any dropped or spilled food and for wiping your hands and mouth. A napkin larger than 26" square is called a lapkin. Napkins almost always came in sets of four, six, eight, twelve, eighteen, or twenty-four, and usually had matching tablecloths. If you are fortunate enough to have family heirloom napkins, not necessarily in matching sets, be creative—mix and match them, but do use them. Your ancestors want to be remembered, and your family and guests will delight in using the napkins.

Napkin

Here are three sets of white linen napkins, from three different European countries, that can be used for breakfast, lunch, brunch, or tea, or even as cocktail napkins. Top left: Lace edged, from England, c. 1920s. Set of six $18-24. Top right: From Madeira, with white-on-white cutwork and embroidery. c. 1940s. Set of four $12-16. Bottom: Set of six white linen napkins with cutwork, light blue thread embroidery, and Filet Lace edge. Italian, c. 1940s. $30-36. The value of the Italian napkins is higher because they have never been used, they retain their original paper label of origin, and they have extensive handwork and lace edging.

A bolt of twelve Irish damask linen napkins. The scissors indicate
the pulled thread mark on which to separate the napkins for
cutting and hemming. 24" square. c. 1920s. $48-60.

Napkin Rings - When one speaks of napkin rings, what comes to mind are metal bands embossed with monograms. The sole purpose of a napkin ring was to differentiate one's own napkin used for breakfast, lunch, and dinner for many days in a row from some-one else's napkin, also used for many days in a row. This is the reason why napkin rings from years ago were never in matching sets. They were considered personal items and were often taken along when one moved away from home.

Napkin Rings

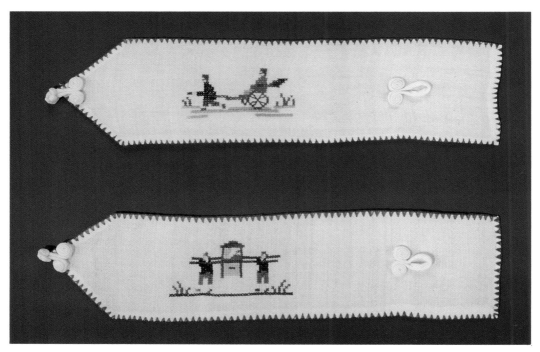

These two napkin rings are made of rice linen with an Oriental design, that includes
the frog closures. Cloth napkin rings are most unusual. These came from a set of six,
the other four are unfinished. c. 1940s. $12-15 for the two finished rings.

Needle Lace - First developed in Venice, this lace is worked entirely with a needle and thread over a pattern, instead of over a permanent foundation, such as net. A pattern is drawn on parchment, then a thread is pinned on the pattern and the decorative loops are stitched to this thread. Forms of needle lace are Alençon, Venice, Point de France, Point de gaze, Reticella, Rose point, and cutwork.

Needle Lace

An Italian Needle Lace tablecloth with baskets of fruit and figures, from about 1900-1915. Made of linen thread, this *point de Venise* cloth is very desirable. 124"L x 72"W. $650-700.

Net - This is a square mesh background on which laces such as filet net or darning can be manipulated. Some open construction net can be created by weaving, knitting, knotting, or other methods.

Net Darning - Some of the first articles of net darning known to man were fishnets. On the smaller and fancier side is net darning, also known as filet darning. It consists of a net background (either hand-made or machine made) with simple darning stitches woven back and forth, over and under.

Net Darning

This is a piece of handmade net background with a figure made of darned net. To create the figure, a needle was used to weave stitches back and forth, over and under. Although it was purchased in France, the origin of this piece is unknown. 18"L x 12"W. $20-30.

These two lovely net darning runners came from the same estate and were originally purchased together in New York City in 1925. The oval runner has a swirling pattern with lilies as the main motif. 38"L x 18"W. $25-30. The rectangular runner also has lilies, but its main theme is butterflies. 42"L x 18"W. $25-30.

Although handmade net darning is still made in Europe, machine-made pieces without much skill or design are manufactured in Asia. These two doilies are examples of such poorly made pieces from Asia. c. 1950s. $5-8.

This Italian ivory bed cover has needle woven figures in two tones of thread. The contrast
is quite striking. This bed cover was from a 1920s trousseau that was never used. $250-300.

New, Old Stock - This term refers to old linens that have never been used. Such linens are identified by the original paper labels and price tags still attached, or by old pencil-written sizes or prices. Quite often, a trousseau will include most of the linens used during the owner's lifetime and a few never used pieces kept for a "special occasion" that never came. Fortunately for us, these treasured pieces can now be enjoyed and will surely last our lifetime. If properly used and cared for, they too will become heirlooms for our own descendants.

New, Old Stock

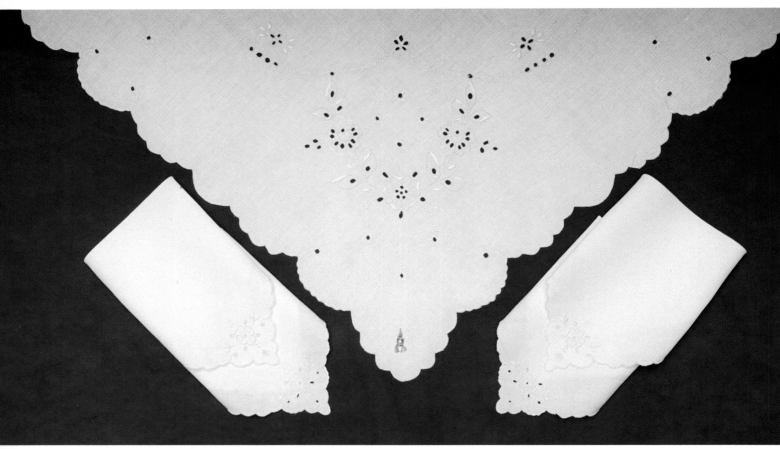

Above:
A white linen Madeira tea cloth with hand cutwork and embroidery, plus six matching napkins. This old, but never used, tea cloth retains its metal tag bearing the name of the maker. c. 1940s. $35-40.

Left:
A close-up of the maker's metal tag.

Top left: A damask woven guest towel with a yellow band of color, still bearing its "Made in Czechoslovakia" paper label. c. 1930-1940. $20-25. Top center: A linen runner with a "Made in Madeira" paper label. c. 1920-1930. $25-30. Top right: An Irish linen damask tablecloth with a descriptive paper label. c. 1940. $35-40. Bottom right: A pair of Italian cotton cutwork pillowcases with a "Made in Italy" stamp. c. 1940. $25-30. Bottom center: A linen 36" square tea cloth with a "Made in Azores" paper label. c. 1940. $25-30.

Here is an assortment of linen kitchen tablecloths from the 1940s that have their paper labels still attached. $20-30.

Normandy Lace - Assembled in Normandy, France, these very highly priced pieces are strips of machine-made Paris Lace hand sewn together with pieces of hand embroidered or machine embroidered centers of fabric. It was fashionable for American soldiers returning from war to bring home such pretty Normandy Lace items to a loved one. Most pieces found are bedspreads, doilies, runners, boudoir pillows, and tablecloths.

Normandy Lace

Normandy Lace is also referred to as patchwork lace because of the strips of Paris Lace, medallions of hand embroidered fabric, and Net Filet Lace edging. This piece is 24"L x 18"W. France. c. 1910. $50-75.

Opposite page top:
This doily is very busy because of the many different types of laces. There is Cluny Lace, Paris Lace, Irish Crochet Lace, Needle Lace, Net Filet Lace, Bobbin Lace, and embroidered linen. If there were room for the kitchen sink, it surely would be there too! 30"L x 20"W. French. c. 1920s. $35-45.

Opposite page bottom:
Another Normandy Lace doily.

Organdy - A lightweight sheer cotton, organdy has an acid finish to give it the stiffness needed for a crisp look and feel. Due to its stiffness, organdy tends to wrinkle easily. Placemats, tablecloths, and runners are the preferred linens made from organdy.

Organdy

This placemat is from a set of eight. It has an appliquéd border and flowers, with hand embroidered vines. c. 1920s. Set $45-55.

A glass coaster with peach colored calla lilies and leaves appliquéd onto the organdy. 6" round. Set of eight $42-48.

Overshot Coverlets - Also called a "linsey-woolsey" because of the linen and wool weave, these were woven on small portable looms, usually by traveling weavers looking for room and board in exchange for weaving a coverlet. The "overshot" pattern was the most frequently used pattern, thus giving this form of Early American coverlet its name. The linen was always a natural color and the wool usually a dark color, such as indigo blue, burgundy red, or forest green. In the summer the cooler linen side was used and in the winter the darker wool side was used.

Overshot Coverlets

A typical indigo blue wool and natural linen overshot coverlet with a linen fringe. Made in New England in a common pattern. c. 1870. $300-400. *Courtesy of Mimi Duphily.*

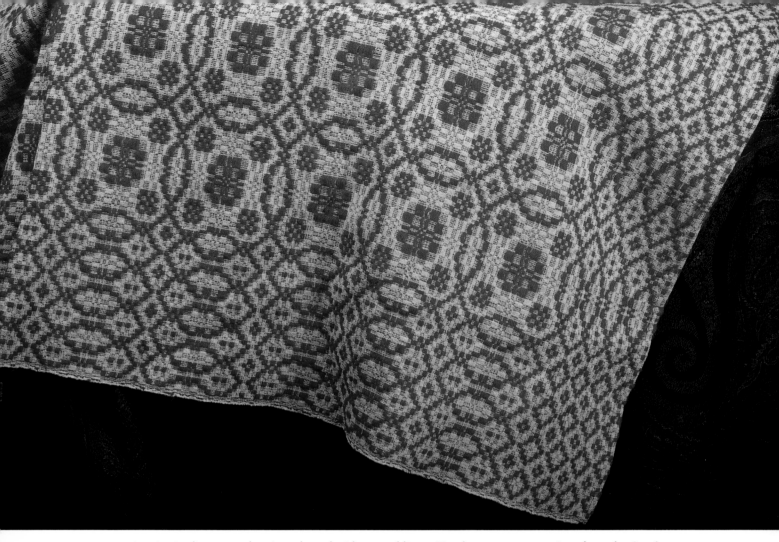

A twin size linsey-woolsey in red wool with natural linen. Hand woven on an antique loom by Carol Tripp, Lakeville, Massachusetts. c. 1980. Whig Rose Pattern. $250-300. *Courtesy of Mimi Duphily.*

Opposite page:
This padded satin stitched tablecloth is a very fine example of white-on-white embroidery, also called French embroidery. The high relief in the stitching shows that this piece was sewn by a very skilled needle worker. French, c. 1910-1920. 66" square. $100-125.

Padded Satin Stitch - This exquisite but tedious embroidery stitch has a very raised, smooth satin look. Used extensively for monograms, the threads are very closely stitched together, never overlapping.

Paris Lace - Also called Point de Paris, this lace originally started out as a handmade pillow lace because it was never a wide lace. With the 1883 invention of a lace making machine in Switzerland, this handmade lace soon fell out of fashion. Normandy Lace is one of the finest examples of Paris Lace still found on the vintage linen market today.

Percale - Of all cotton weaves made, percale has the highest count of long staple threads available: 180-280 thread count per square inch. Higher counts of percale (up to as much as 300 threads per square inch) are also available, but remember: the higher the count, the thinner the threads, and the weaker, but softer, the fabric. Whenever linen for bedding was not available, only the best percale cotton was used instead.

Padded Satin Stitch

Pillow Cover - Also called cushion covers, these ornamental covers were highly decorative, often works of art. Pillow covers were designed to be seen and used; therefore they often contained the fanciest stitches worked in great detail. Usually meant for the bedroom, pillow covers also complement the parlor as cushions for side chairs, settees, loveseats, and sofas.

Pillow Lace - Considered a form of weaving, pillow lace is made on a pillow with bobbins that have linen thread wound around them. Often there can be anywhere from a dozen bobbins to as many as several hundred.

Pillow Cover

A trio of pillow covers, or pillowcases. The top cover is often called a pillow sham because of the button closure on the back. The second cover is a pillowcase, and the third cover is also a pillowcase with a lace edging. c. 1920s. 16"L x 12"W. Each $20-25.

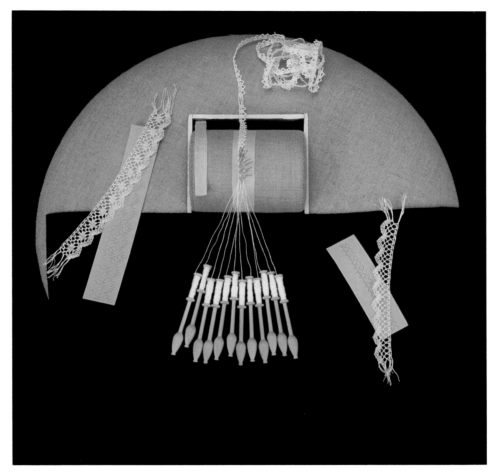

This round pillow shows bobbin lace being made. Two swatches of finished lace are to the left and right of the work in progress. *Courtesy of Mimi Duphily.*

A close-up showing how the straight pins hold the linen threads in place for weaving while on the parchment pattern. *Courtesy of Mimi Duphily.*

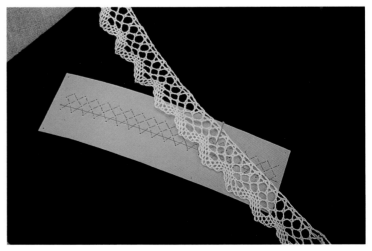

Another close-up of handmade bobbin lace with the parchment paper pattern. *Courtesy of Mimi Duphily.*

Pillow Shams - As the name implies, a sham means a cover-up for a pillow. Shams come in many different sizes, shapes, and forms. The European sham is oversized, about 26" to 32" square. The American sham is standard pillow size, about 26" to 30", in a rectangular form. A layover sham is one that you place over your wrinkled pillowcases just for the day, removing it at night. At one time, people could afford only one pair of pillows or one long bolster, thus the need for a layover sham. Some shams have buttons or ties on the back to secure the pillows within.

Pillow Shams

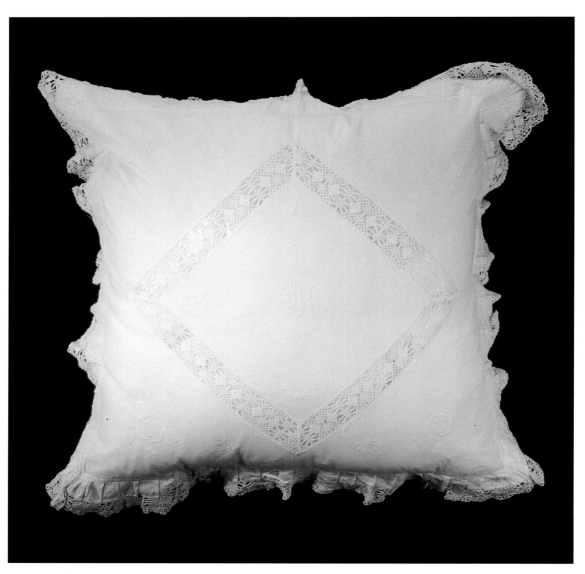

This German white embroidered cotton pillow sham with machine-made lace inserts is 26" square and has a button closure on the back. Such shams are used as pillowcases in Germany. The front is for show and the back is for sleeping. c. 1940. Pair $100-125.

Next page:
A pair of square lay-over shams with red turkey work embroidery. c. 1890s. $45-50.

Pillowcases - Often called a pillowslip, this is a traditional everyday household item. Pillowcases can be of any size, even custom made, to fit any size pillow. The most sought after case is one made of linen, because it is so cool in the summer and warm in the winter. The standard size of American pillows is 26" long x 20" wide. Queen size is 30" long x 20" wide, and king size is 36" long x 20" wide. Vintage cases can be oversized to fit our modern queen and king size pillows simply because the fancy cutwork and embroidery on the ends would mark one's face, thus extra length was added. Since sheets were turned down quite a bit, the extra length of the pillowcase would also cover the side view of the mattress and boxspring.

Pillowcase

A pillowcase sewn on three sides with the fourth side unsewn to accommodate the pillow. This has cutwork, embroidery, and a lace edge. c. 1940s. 34"L x 20"W. Pair $15-25.

A pair of linen pillowcases with mosaic work. These cases are extra long so one's face would not touch the handwork (which would cause marks on the face). c. 1920s. 36"L x 20"W. Pair $45-60.

Pincushion - When handwork was in fashion, one would usually have more than one pincushion. The most popular was the red tomato, which held your pins with the attached strawberry that was filled with a very fine pumice for sharpening the needles. The other type of common pincushion had a hard sawdust filling with a two-piece white-on-white fancy embroidered cover.

Pin Cushion

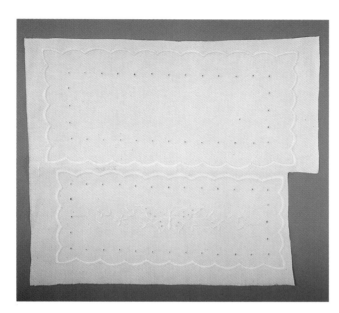

Left:
A pin cushion, made from a kit, that was hand embroidered but never cut apart.c. 1900-1920. $12-15.

Below:
These three finished pin cushions were made from kits. The cushion inside is either a hard rubber material or tightly packed sawdust. These are prettier than a red tomato pin cushion and were used during sewing bees. Today, brooches and stick pins adorn these pin cushions rather than straight pins. Top left: White-on-white embroidered pincushion with a lavender ribbon. c. 1900. $25-30. Right center: Printed silk fabric. c. 1890. $25-30. Bottom center: White-on-white embroidered with a green ribbon. c. 1900. $20-25.

Placemat - To take the place of a tablecloth, individual placemats were sometimes used. A standard dinner set included a center table runner, matching napkins, and at least twelve placemats. Luncheon sets were slightly smaller and consisted of four to twelve settings.

Point Lace - A type of Venetian Lace, also referred to as Rose Point Lace, this is a needle lace worked on a delicate net background on which elaborate designs have been worked. It is similar to early gros point in design, but with a less heavy outline. Most Point Lace came from Italy, quite possibly from the highly skilled hands of nuns.

Placemats

This placemat is one of twelve embroidered with light blue thread. The cutwork was done by hand on the island of Madeira. c. 1940s. $70-80.

One of a set of eight placemats made of organdy with a fine cotton appliqué for the border and flowers. The napkin is made of linen, for absorption. c. 1940s. $80-90.

A complete set of eight placemats, eight napkins, and one center runner.
This linen set was handmade in Madeira and has appliquéd calla lilies
and leaves, with hand embroidery for the stems. c. 1940s. $80-90.

Quaker Lace

Quaker Lace - To imitate beautiful handmade lace without the exorbitant price, the Quaker Lace Company began producing some exquisite machine-made cotton cloths for middle class households sometime in the 1940s. Today, the company's laces are a combination of acrylic blends.

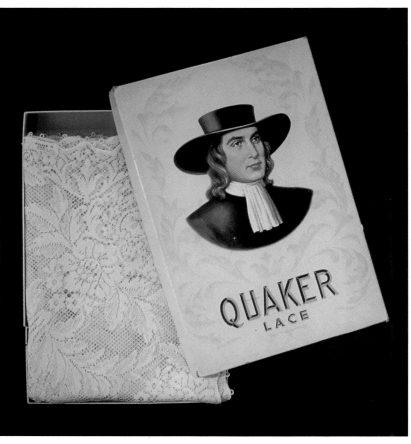

The name Quaker Lace has been a household word since the company's beginning in the 1940s. Some fancy tablecloths made by the company have figurals and beautiful scroll-work. This piece is 104"L x 72"W. Ivory cotton. $75-100. *Courtesy of Jeanne B. Clarke.*

Rayon - While trying to make synthetic silk, developers manufactured rayon instead. The vibrant, lustrous fiber was first used for table linens in the late 1930s; most of these were made in Japan. At the same time, France was making high quality linen and rayon blend tablecloth sets with elaborate damask weaves and lavish patterns. It is fairly easy to remove stains from and to iron the blend fabrics; with all rayon linens, however, food spots are difficult to get out and ironing the fabric is more difficult.

Reticella Lace - The ancestor of all needle lace, this is a technique of drawnwork with a skeleton of woven fabric. Making Reticella Lace involved many steps, so the finished piece often took a number of differently skilled lacemakers to complete. The secret instructions for this lace were highly guarded in the event that a worker might leave the district and take his knowledge with him, an act which was punishable by death.

Reticella Lace

One of a pair of fancy Italian linen and embroidery guest towels with Reticella Lace, a needle lace. c. 1920. $35-40.

A close-up showing one of a pair of guest towels with figural Reticella Lace. 24"L x 14"W. Italian. c. 1920. $40-50. *Collection of the author.*

Ribbon Embroidery - Rare, exquisite, and highly sought after is silk ribbon embroidery. In the many years I have collected linens, the only such embroidery I ever purchased was in France. While staying in a Paris apartment bed and breakfast, I was fortunate enough to buy the woman innkeeper's complete trousseau. The "maiden" at 58 years old had never married. Two examples I have seen of this work were embroidered on moiré, one was worked on silk. Silk ribbon embroidery is easy to make, yet it never became a popular form of embroidery to which all honest women were expected to devote themselves from childhood.

Ribbon Embroidery

Close-up of a lambrequin with wonderful silk ribbon embroidery on moiré silk. Silk ribbon embroidery is easy to make yet few pieces are ever seen.

Rice Linen - Almost always made in China, this coarse, often brittle linen does not have a long lifespan as it tends to break more than finer woven linen and has fibers that separate very easily. Most rice pieces are not worthy of the wonderful workmanship the Chinese put into the fabric, and would typically have been purchased in 5¢ and 10¢ stores in the 1940s. The most popular patterns have Oriental designs of chrysanthemums, dragons, and temples. Also known as grasscloth, rice linen is made from a tall perennial shrub from the nettle family that requires a hot, humid climate. It is one of the strongest natural fibers and its strength increases when it is wet. It has a silken luster although it is stiff and brittle and lacks resiliency.

Rice Linen

A rice linen tea cloth from China, with beautiful hand embroidery and very well executed drawnwork. This is a fine example of rice linen and a good addition to any collection. c. 1940s. 48" square. $25-35.

This rice linen bureau scarf has hand-embroidered Oriental chrysanthemums and hand stitched drawnwork. c. 1940s. 46"L x 16"W. $25-35.

Rickrack

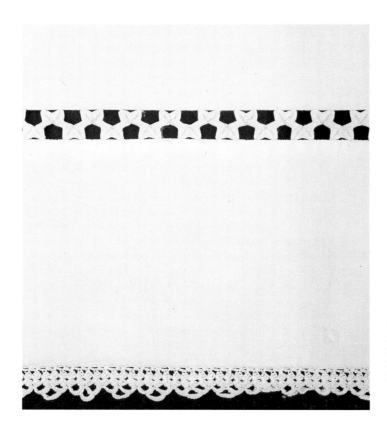

Rickrack - The very first machine-made novelty tape, rickrack resembles a zigzag stitch and is as popular now as it was in the early 1900s when first manufactured. By using this tape, the desired piece could be finished very quickly. The rows of rickrack were most often attached together with embroidery stitches.

A close-up of handmade rickrack. Rickrack was made by turning and folding a premade tape in any desired shape possible.

Runner - Meant for rectangular-shaped furniture such as dressers, chests, sideboards, or buffets, runners were typically made in the same rectangular shape as the furniture. They can be found also in other shapes, as well as in any fabric, lace, or size available. They were meant to protect the finish of wood and the surface of marble-topped furniture. Some runners can also be used on tables that are extended with leaves to become rectangular in shape. A runner with only three sides that has a design or decoration similar to a lambrequin, however, was never meant to be used on a dining table.

Runner

Top: A linen Madeira bureau scarf or table runner with hand cutwork and a very scalloped edge. c. 1920s. 52"L x 18"W. $25-35. Bottom: A very simple yet elegant linen runner with handmade bobbin lace edging. c. 1910. 68"L x 17"W. $20-30

A linen centered runner with machine-made mosaic work and a pleasing design. Notice the two side seams attaching the work together. c. 1940s. 54"L x 16"W. $20-30.

A very detailed linen runner with cutwork, embroidery, and a scalloped edge. 72"L x 18"W. Madeira, c. 1920s. $35-40.

Rust - To remove old rust spots, I recommend Rit Dye Rust Remover, available at hardware, craft, and grocery stores. Using lemon and salt leaves a yellow spot from the lemon juice, plus the salt is too abrasive for the already weakened rust spot area. Whink is pure hydrofluoric acid and is too harsh to use on fine fabrics and lace.

Sachet - Known as a keeper for potpourri, the purpose of a sachet is to keep moth retardants, such as dried lavender flowers, dried rosemary, and wormwood, in a fabric holder among one's fine linens, laces, and stored-away fabrics. Although perfume is also used to lightly scent fine lingerie, perfume will stain most fabrics.

Sachet

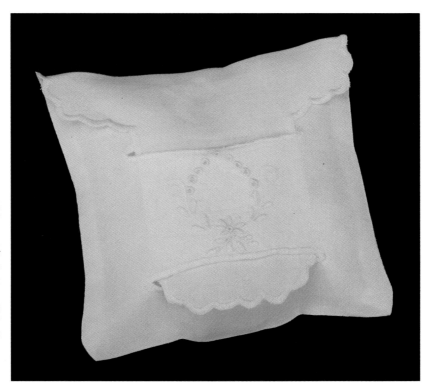

This linen sachet bag with hand embroidery had dried rosemary and lavender flowers in it when purchased. If rosemary and lavender are unavailable, just put some perfume on the enclosed fiberfill lining to lightly scent your drawers. c. 1900. 6"H x 4"W. $10-12.

Sampler - Early samplers were made by young girls as works of art to demonstrate their skill with a needle. In order to mark the household linens and clothing, it was necessary to master the skill of fine and fancy needlework. The samplers were patterns of stitches and designs to be used after the girls married as their reference for marking the household linens.

Sampler

Four different styles of lettering are shown in this sampler, which was purchased at a yard sale in Paris. It is dated 1907 and is incomplete. 10"H x 12"W. $75-100.

Shadow Work - This form of embroidery is worked on the back side of the fabric rather than the front, to give the illusion of a shadow from behind. The work is then outlined on the front side of the fabric with either a buttonhole stitch or a heavy outlining running stitch.

Shadow Work

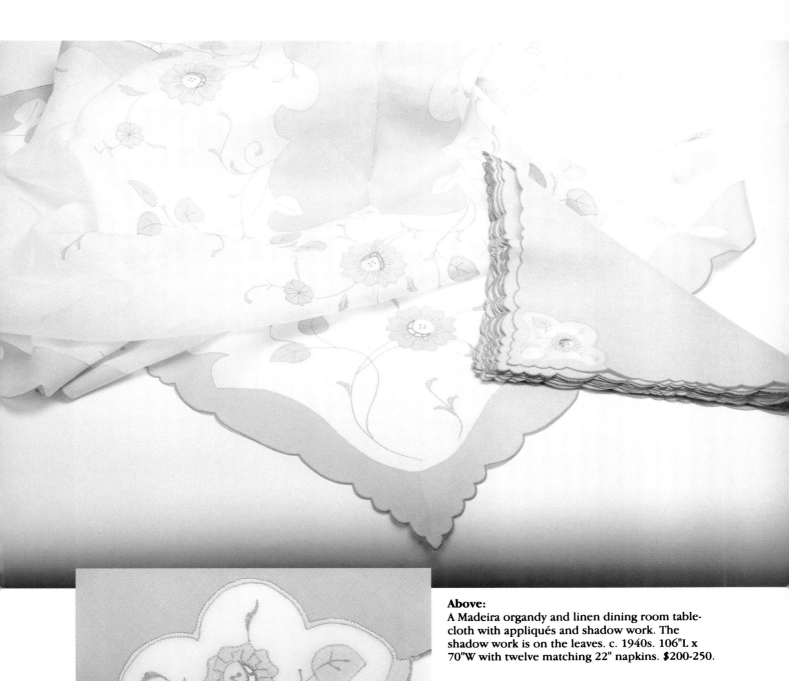

Above:
A Madeira organdy and linen dining room table-cloth with appliqués and shadow work. The shadow work is on the leaves. c. 1940s. 106"L x 70"W with twelve matching 22" napkins. $200-250.

Left:
A close-up of the shadow work leaves from the back side, showing the figure eight crossover of embroidery stitches.

Sheets - An important household staple, usually dozens of sheets were found in early trousseaux. Bottom sheets began as flat sheets and were meant to cover the mattress ticking. Then, in the 1960s, the fitted bottom sheet became an instant success, hailed as a time saver for the tedious daily chore of bedmaking. The top sheet was usually oversized and designed to protect the coveted blanket from wear, tear, and soil. The most desirable and sought after linen top sheets have lavish cutwork, embroidery, and lace, and are large enough to fit our modern king and queen size beds; their fold over of 10" to 25" shows off the design and protects the blanket edge. Yes, king and queen size beds do exist in those enormous stone castles in Europe and do require oversized bedding.

Sheets

Above:
This cotton trousseau sheet from Italy shows drawnwork lace inserts, hand embroidery, an inverted pointed edge, and Filet Crochet Lace. This is just the corner, the center is much more elaborate. Sheets can be as plain or as fancy as desired. c. 1900. 98"W x 108"L. $75-95.

Left:
This sheet set had a paper label indicating a 1929 copyright. The sheet and pillowcases were factory made, while the floral border was hand sewn at a later date. c. 1929. Cotton. $45-60.

Silk Embroidery - The Chinese silk fabric industry began during the twenty-seventh century B.C. when silk filament, sometimes one mile long, was unwound from silkworm cocoons. Silk embroidery was a practiced skill all during the 1800s but faded rather quickly when less expensive cotton embroidery floss became available around 1900.

Silk Embroidery

A very fancy linen centered runner showing some beautiful silk hand embroidery. The lace edge is needle work. 62"L x 17"W. European. $65-75.

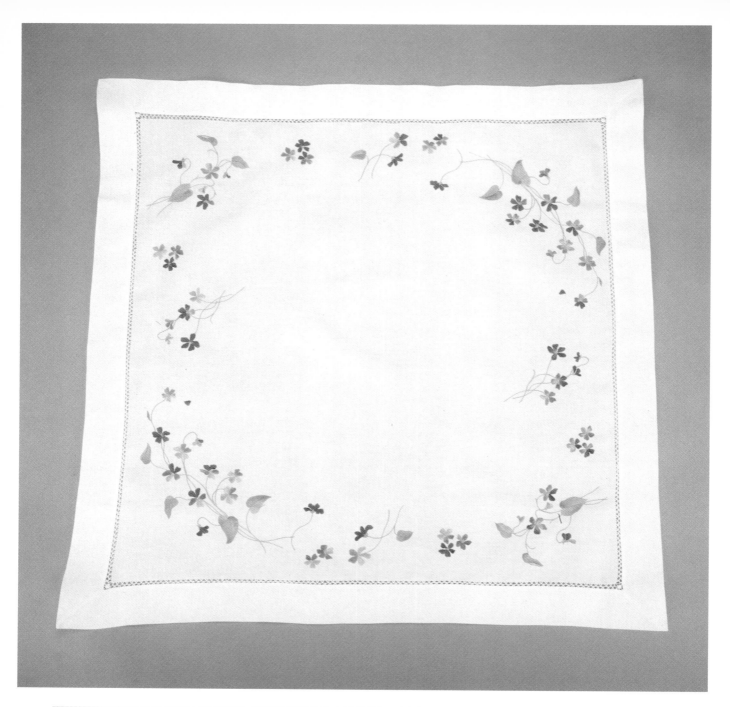

Above:
A linen tea cloth with an interesting ring of silk embroidered violets. c. 1900. 36" square. $45-55.

Left:
Close-up of the silk embroidery variation.

Spain - Spanish needle lace was strongly influenced by the Punto in Aria style of Italy, but also incorporated Moorish geometric designs using gold and silver threads. The legacy of Spanish handwork is decorations with scalloped edges, baskets of flowers, and ribbons.

Spain

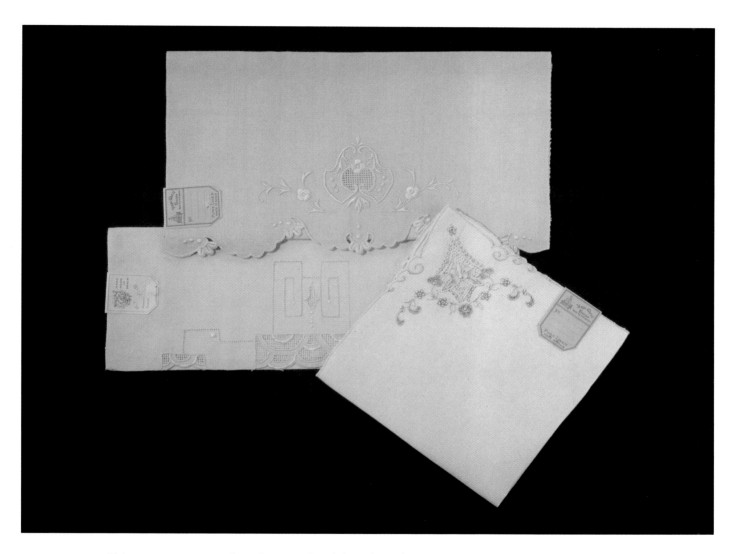

If these two guest towels and one napkin did not have their original paper labels, I would have thought they came from either Italy or Madeira. Since Spain, Portugal, and Italy are close together, some of the handwork from each country overflowed into the others, making it sometimes quite impossible to tell the difference. Cream linen guest towels, each $10-12. The napkin is from a set of twelve with a matching tablecloth that is 100"L x 68"W. $200-250 for the set.

Splasher - A splasher's only purpose was to be attached to a wall or hung over the high bar of a commode to prevent water from being splashed on the painted or papered wall as a person washed up at the commode from a pitcher and bowl set. Most examples found are 20" x 30" and have the maker's monogram exquisitely executed, providing a chance to admire her best handwork.

Starch - Potato starch was invented purposely for the appealing look and feel of crisp linens, for increased body, and for abrasion resistance. Those are the pros—now the con: it attracts silverfish bugs that thrive on potato starch. Fortunately, modern starch does not contain any food products that attract insects. If you do want crisp linens, starch them lightly prior to use and avoid ironing any creases. However, linens that are to be stored should never be starched.

Splasher

The people from whom I bought this linen piece told me that it hung over a commode towel bar and was used as a wall guard or splasher. I believe its real use was as a baby's crib cover, but one makes do with what's available. c. 1920s. 36"L x 24"W. $25-30.

This linen splasher with mosaic work was proudly displayed on the back of a guest bedroom door. The owner placed it there, rather than on a wall above a pitcher and bowl set, so that it wouldn't be used too often. She said she made it in 1929 when she was ten years old and wanted it to last a lifetime, which it did. 60"L x 24"W. $35-45.

Starch

The two questions most asked of a linen merchant are: "How do you get your linens so white?" and "How do you get your linens so crisp?" I tell them "Soak, soak, soak" and "Starch." Most people have never heard of liquid starch and are very impressed with the beautiful finish it gives. Linit Starch is white in color, Sta-Flo Starch is colored blue. The complete directions are on the back of the bottle. Starch is available at most grocery and craft supply stores.

Storage - Rotating your timeless treasures is very important. When they are not to be used, the spotlessly clean, completely rinsed, and thoroughly dried linens should be put away unironed, unstarched, and unfolded. They should be rolled up, wrapped in acid-free tissue paper, and enclosed in archival storage boxes. If you really treasure your prized possessions, the above-mentioned steps will be easy to follow. Your local museum can give you the name of a museum supplier from whom you can obtain acid-free paper and boxes. Make the effort. Your descendants will cherish your collection as much as you because you took the time to properly preserve their future heirlooms.

Stuffed Work - Mostly used on quilts, these designs were formed by piercing the fabric on the back, stuffing in cotton, and working the hole closed by the head of a needle. The Italian form of this work is called Trapunto and the French form is called Marseille.

Swedish Embroidery - Quite possibly from Sweden, this type of hand embroidery was used only on the top side of huck linen. The counted stitches were never seen on the back side of the linen. Typically, colored embroidery floss was used mainly on towels.

Swedish Embroidery

A pair of hand towels on huck linen with pulled Swedish Embroidery. c. 1920s. $12-15.

Swiss Embroidery - Not only did the finest hand embroidery come from the Swiss hamlet of Appenzell, the finest hand-guided machine embroidery also was produced in Switzerland. Known as tambour, it is basically a chain stitch, a very simple embroidery stitch that is pleasing to the eye because of the flowing roundness of the work.

Swiss Embroidery

A close-up of the paper label on the blue runner.

A light blue cotton runner with Swiss Embroidery and a beautiful paper label. Machine stitched. c. 1950s. 34"L x 16"W. $14-18.

Tablecloths - In country homes, tablecloths were used to cover the one and only scrubbed-top worktable. In city homes, tablecloths were used on fancy lacquer finished tables. In addition, cloths were used for anything from 36" square card tables to banquet tables large enough to seat 100 people. They can be square, round, rectangular, oblong, or oval. Interior photographs of Victorian dining rooms showed rectangular dining tables draped in rectangular white damask cloths large enough for the four corners to puddle on the floor. The oval dining cloth is a 1950s invention which doesn't have the elegance or the draping effect of a rectangular cloth, and thus is not as desirable.

Tambour - This type of chain stitch hand embroidery is made with a tambour hook on a frame with fine muslin or net. Machine-made tambour is either called Swiss embroidery or Limerick Lace from Ireland.

Tablecloths

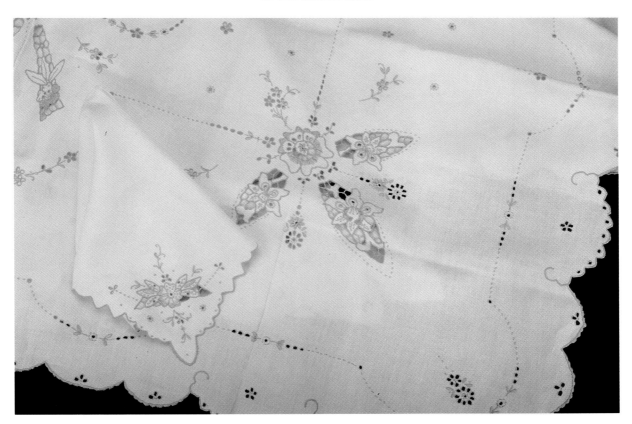

An ivory linen tablecloth made in Portugal with twelve matching napkins that are 22" square. This color of linen was used from 1940 on. Dishes were becoming cream colored instead of stark white so the ivory color and colored damask were all the rage. c. 1940s. 124"L x 70"W. $125-150.

Tambour

A tambour work poodle that was stitched by machine. The work is guided by hand, however, not by computer. 16"L x 12"W. c. 1940s. $10-12.

This is a close-up of a curtain panel showing handmade tambour work. Machine-made tambour work is called Swiss Embroidery or Irish Limerick Lace. 106"L x 42"W. Pair $80-100.

Tatting - Made with thread wound on an oval shaped two piece shuttle, tatting dates from the eighteenth century, quite possibly from Italy. Used mostly as edging and trimming, it uses one basic knot to secure loops and picots.

Tatting

A pair of percale cotton pillowcases with a hand tatted lace edge. c. 1920. $25-35.

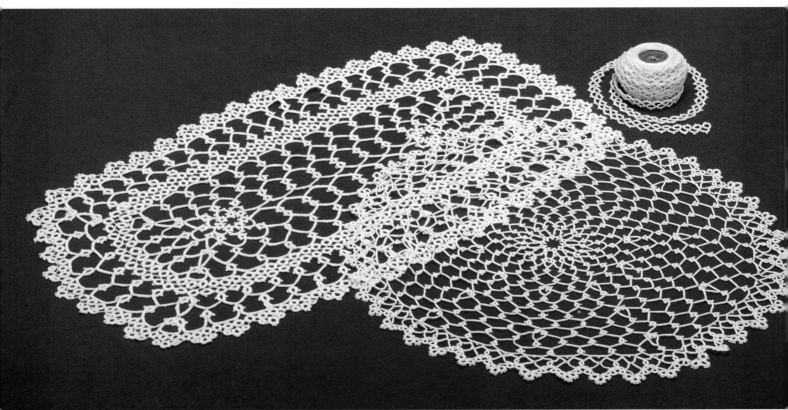

These two tatted lace doilies were made by the same lady, Ruby Westgate Hall, of Brockton, Massachusetts, in 1924. Each $20-25.

Tea Cloths - Whenever high tea is served in Great Britain at 5:00 in the afternoon, the finest silver, china, and linen are used to celebrate this daily event. Using tea cloths during this time is a way for the hostess to show off her exquisite handwork embellished with the finest adornments possible. Today, these beautiful square cloths are also used as table toppers in the bedroom or front parlor.

Tea Cloths

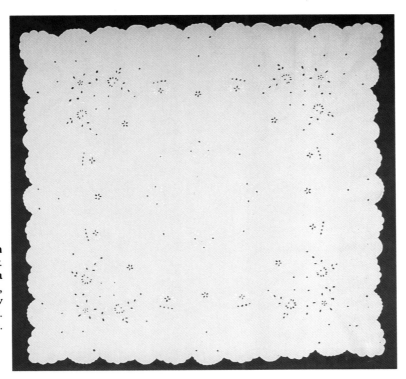

A Madeira linen tea cloth with handcut embroidery. Since we no longer use tea tables for afternoon tea, cloths such as this are now used as table toppers. c. 1940. 36" square. $25-35.

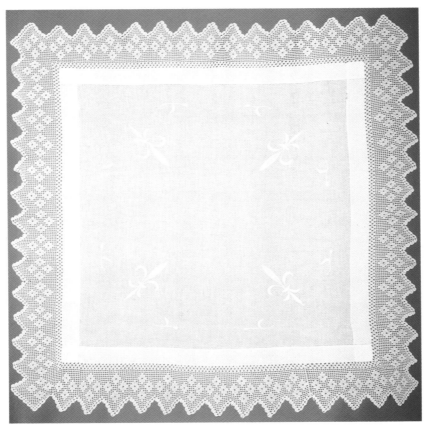

A true English tea cloth with machine embroidered fleur-de-lis designs on cotton and a hand crocheted filet lace edge. c. 1920s. 48" square. $75-100.

Tea Cozy - These two-sided covers have a feather, flannel, or cotton batting filled liner and are meant to keep the tea pot hot. Most cozies I have found came from England, where taking tea is an art.

Tea Tray Cloths - Customarily, silver tea sets have trays on which the pieces are placed. In order to prevent the silver-footed hollowware from scratching the tray, oval or rectangular cloths were placed between the tray and the silver. These cloths can be tailored damask or extremely fancy.

An English cotton crocheted tea cozy with a feather filled liner to keep the pot warm. c. 1920s. 12"H x 10"W. $35-45.

Tea Cozy

Tea Tray Cloths

First, from top: An Italian cotton tea tray cloth with cutwork. Second: An Irish damask linen cloth with a hemstitched edge and original paper label. Third: An Italian ivory linen cloth with a Reticella Lace insertion. Fourth: A French Cluny Lace cloth in linen. c. 1920-1940. Each $20-30.

Tenerife Lace - Originating in the Canary Islands off the coast of Africa, this lace is made on a round frame with pins around the edge. A needle and thread are used to form the pattern. Also called wheel lace.

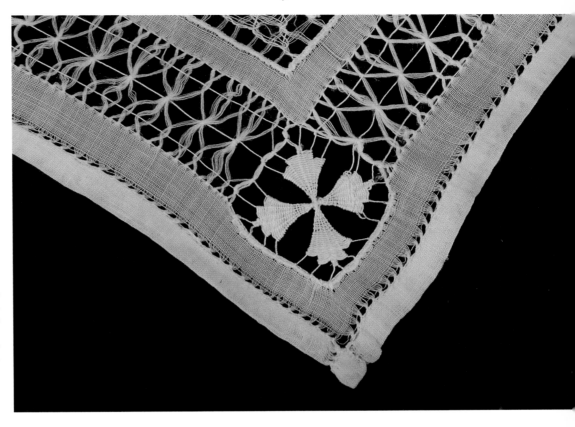

Close-up of a table square with beautiful counted drawnwork and a Tenerife motif in the very corner. Tenerife Lace originated in the Canary Islands.

Ticking

Ticking - This was originally handwoven in linen as a covering for feather mattresses and pillows. Its characteristic herringbone pattern was meant to keep the feathers in and the ticks out!

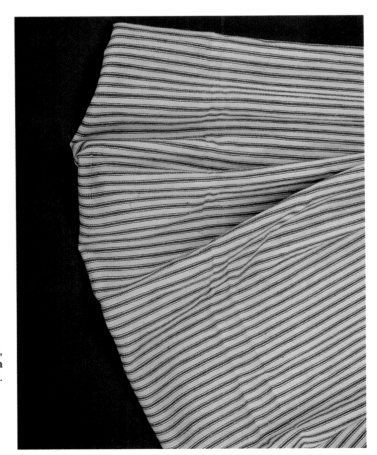

A fabric familiar to us all, ticking is still made but in lighter weights of cotton.

Torchon Lace - French for rag, Torchon is a strong bobbin lace of continuous linen thread laces with geometric designs. Like its close cousin, Cluny Lace, Torchon has spiders, fans, and leaves. Since 1900, machine-made Torchon and Cluny Lace have been hard to tell apart from handmade pieces.

Torchon Lace

A linen centered Torchon Lace runner. French, c. 1900-1920. 56"L x 18"W. $65-75.

This French Torchon Lace round linen center doily has a rampant lion design in the woven lace. 36" round. c. 1900. $50-75.

Towels - Used in either the kitchen, bathroom, or bedroom, towels come in many different sizes and shapes. Linen was the most desired fabric and huck the preferred weave. Linen fibers are lint free, hence their continued use in kitchen towels. Huck weave is a small nubby fabric and is highly absorbent. Fingertip towels were found mainly in guest bathrooms and bedrooms, along with hand towels and body towels. Fine woven damask patterned towels often have monograms, fancy handwork, and lace trimmed edges. The fancier the towels, the more highly sought after they are today for home decorating and gift giving.

Towels

Above:
Top left: A huck linen hand towel with Madeira hand cutwork. c. 1920s. $10-12.
Top right: A green linen hand towel with compass work and embroidery. C. 1940s. $8-10. Bottom: An Austrian mosaic work linen bath towel. c. 1930s. $15-18.

Left:
Four assorted hand or guest towels with hand embellished cutwork, lace, and fine embroidery. Each $15-20.

Four Italian linen damask towels with hand knotting and fringe. Two are embroidered with the original owner's name. 60"L x 30"W. Each $25-40.

Trapunto - Italian in origin, this cotton white-on-white design is machine made and has the appearance of stuffed work. Although machine stamped, not hand stuffed, Trapunto style coverlets are very desirable and duplicate the look of a wedding quilt without the same high price. The trapunto technique is also used for items such as baby bibs and petticoats.

Trapunto

Close-up of a white-on-white Trapunto style Marseilles coverlet. The machine-stamped pattern is in high relief. c. 1880-1920. 88"L x 84"W. $150-200.

Close-up of another coverlet.

This bib is machine
stitched with a fine
cotton filling. c. 1880.
$20-25.

Trousseau - At the beginning of the twentieth century, tradition dictated that a trousseau included the bride's wardrobe, bedding, and all household linens, including aprons. These were assembled and designed to last a lifetime. Each piece acquired would be painstakingly monogrammed and later lovingly maintained. Some trousseaux contained fifteen pairs of sheets, thirty pillowcases, sixty hand and body towels, three dozen napkins, and one dozen various sized tablecloths. In some families, a woman's trousseau might represent the only real treasure in her dowry. Sewing, embroidery, and weaving, including the trousseau items, were among the needlework to which all honest women were expected to devote themselves from childhood.

Trousseau

One of a pair of linen pillowcases with a bride and groom
pattern from a wedding trousseau. 36"L x 20"W. Italian. $50-60.

Turkey Work - Legend has it that the red dye used in producing red fabrics one hundred years ago came from the brightly colored yolks of turkey eggs; the name "turkey work" evolved as a result.

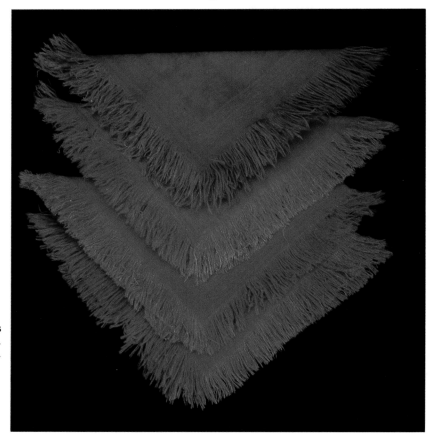

A wonderful set of deep red turkey napkins in a damask pattern with a fringed edge. c. 1880-1900. 16" square. Set of four $40-48.

Bottom: A damask woven linen tray cloth with red embroidered violets and a fringe. c. 1880. 36"L x 18"W. $15-$20. Top left: One of a pair of linen damask tea towels with a shamrock pattern, hand knotted fringe, and a red turkey band of color. 52"L x 26"W. c. 1880-1900. Pair $30-50. Top right: One of a set of six turkey red napkins with a fringed edge. Sets of these red napkins are hard to find. c. 1880-1900. 16" square. Set of six $36-42.

Underplates - Fancy round doilies were placed between the soup bowl or salad plate and the dinner plate to absorb any spillage onto the dinner plate. These occasional underplates were used only in formal place settings. The average size was 9" round; they are used today as placemats for luncheons or afternoon tea.

Underplates

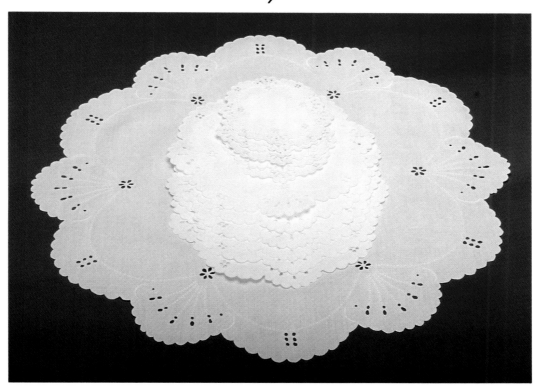

Above:
This dessert or tea set consists of six 5" round undercup doilies, six 8" underplate doilies, and one 26" center doily. Cutwork linen made in Portugal. c. 1930s. $35-45.

Left:
These snowflake pattern crochet doilies are from a set of eight large underplate doilies and eight small underglass doilies. c. 1940s. $35-40.

Vanity Sets - These were three piece matching sets for vanities and deep well dressers. The set consisted of one center piece, perhaps for a vanity tray, and two smaller side pieces for the boudoir lamps. Since vanities are back in bedroom use, the demand for these three piece sets is high and finding all three pieces together would be a lucky find indeed.

Veil Case - As necessary as a handkerchief, a woman's veil was a treasured personal belonging. Linen bags were specially made to protect these lacy triangular adornments.

Vanity Sets

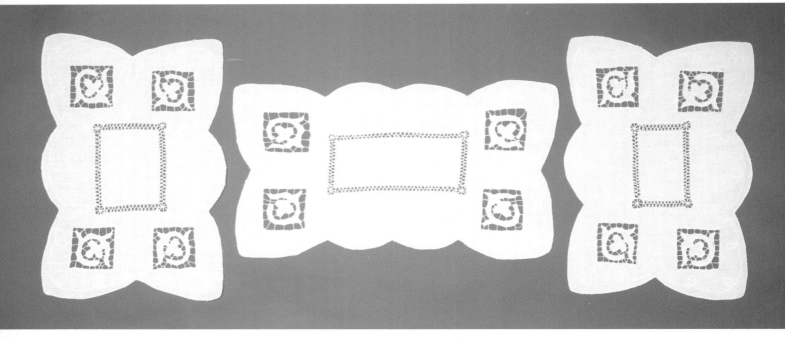

This Italian cotton three-piece vanity set consists of two side doilies to be used for boudoir lamps and a center doily, perhaps for a pin tray. Sets like these were used for deep-well dressers or kidney shaped vanities. c. 1940. $25-35.

A three-piece needle lace vanity set with a linen center. c. 1920s. European. $45-60.

Venetian Lace - A needle lace worked in buttonhole stitch and also known as embroidered lace. The lace is made with a sewing needle and linen thread to resemble "stitches in air" (in Italian, *punto in aria*). This true form of lace was made by filling in a pattern drawn on parchment with many rows of buttonhole stitches with attached brides holding the pattern together.

A round needle lace doily made with a sewing needle and linen thread. c. 1940s. $10-15.

Verse Samplers - During the 1920s, machine-stamped pattern designs with folksy verses and fade-resistant cotton threads were all the rage to easily crosstitch. Not only did this mark the end of free-hand stitched samplers, it also marked the end of hand stitching to mark household linens. In big cities, Chinese and French laundries were readily available to take in laundry and mark it with permanent ink, thus eliminating the need for hand-stitched markings.

Verse Samplers

Verse Samplers for crosstitching came in kit form and were meant to be framed rather than stored away. Unlike samplers of the eighteenth and nineteenth century, they were to be used as guides in later years for monogramming household linens. Verse samplers of the twentieth century were whimsical and had folksy sayings. c. 1940s. 12"L x 10"W. $35-45. *Courtesy of Reneé Maury.*

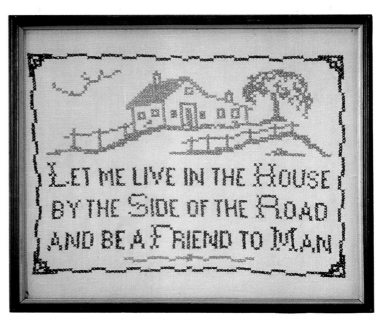

LET ME LIVE IN THE HOUSE BY THE SIDE OF THE ROAD AND BE A FRIEND TO MAN

Wedding Handkerchief - A wedding handkerchief was a very important part of a bride's wedding ensemble. Traditional wedding handkerchiefs were usually passed down from one generation to the next, surviving only because they were used once every 20 years or so. Only the finest European lace was suitable for these handkerchiefs, with more lace than linen being used.

Wedding Handkerchiefs

These four handkerchiefs have handmade tape lace or bobbin lace edges. Three are ivory and one is white. These graceful yet traditional heirlooms have survived for generations because of their once in a person's lifetime use. c. 1880-1900. European. Each $50-75.

Whimsey - A whimsey in glass making was called end-of-the-day glass because using the leftover glass was better than throwing it away. The term whimsey in crocheting was used the same way, and meant using any small bit of leftover thread because dyed or natural batches of thread varied from dye lot to dye lot.

Whitework - Also referred to as white-on-white work, this is white thread embroidery worked on white fabric. Most pieces found today have survived from the 1900s. Earlier examples from the 1800s are too fragile to use, but if made by an ancestor could possibly be professionally framed and preserved for as long as possible.

Zigzag - Net darning lace has a knotted net background with filling threads that zigzag back and forth at sharp angles between vertical threads rather than crossing the design area. Lace curtains made by machine have strong parallel lines with zigzag fillings and shaded patterning.

Whimsey

This crocheted lace tea cup and saucer is strictly a decorative shelf piece, yet requires a skilled hand to accomplish. c. 1940s. $10-15.

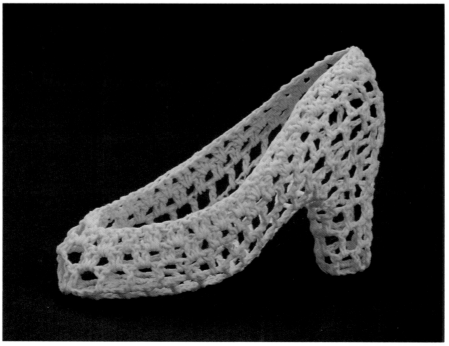

A lady's high-heeled shoe crocheted as a whimsey. c. 1940s. $10-15.

Bibliography

Butterick Publishing Co. *Battenberg and Other Tape Laces.* New York: Dover Publications, Inc., 1988.

deBonneville, Francoise. *The Book of Fine Linen.* Paris: Flammarion, 1994.

Dolan, Maryanne. *Old Lace and Linens Including Crochet, An Identification and Value Guide.* Alabama: Books Americana, Inc., 1989.

Hart, Cynthia and Catherine Calvert. *The Love of Lace.* New York: Workman Publishing, 1992.

Horner, Alda Leake. *Official Price Guide, Linens, Lace, and Other Fabrics.* New York: House of Collectibles, 1991.

Johnson, Frances. *Collecting Antique Linens, Lace, and Needlework, Identification, Restoration, and Prices.* Pennsylvania: Wallace-Homestead, 1991.

Jourdain, Margaret. *Old Lace, A Handbook for Collectors.* London: B.T. Batsford, Ltd., 1988.

Kraatz, Anne. *Lace, History and Fashion.* London: Thames and Hudson, 1989.

Kurella, Elizabeth M. *A Pocket Guide to Valuable Old Lace and Lacy Linens.* Michigan: The Lace Merchant, 1996.

Paine, Melanie. *The Textile Art In Interior Design.* New York: Simon & Schuster, 1990.

Warnick, Kathleen and Shirley Nilsson. *Legacy of Lace, Identifying Collecting, and Preserving American Lace.* New York: Crown Publishers, Inc., 1988.